The Anatomy of
a Sustainable World

The Anatomy of
a Sustainable World

Our Choice Between Climate Change
or System Change

And How You Can Make A Difference

Glen T. Martin

The Institute for Economic Democracy

Institute for Economic Democracy
PO Box 309, Appomattox VA 24522, USA.

Printed in the United States of America
Printing 1.0

Library of Congress Cataloging-in-Publication Data

Martin, Glen T., 1944- author.
    The anatomy of a sustainable world : our choice between climate
change or system change and how you can make a difference / Glen T.
Martin.
          pages cm
    ISBN 978-1-933567-47-1 (pbk. : alk. paper) – ISBN 978-1-933567-48-8
(ebook)
    1. International organization–Environmental aspects. 2. Sustainable
development–Political aspects. 3. Climatic changes–Political aspects. I.
Title.
    JZ1318.M37645 2013
    341.2′1–dc23
                                              2013010205

Contents

Introduction

Human awareness of our environmental and social situation on the Earth today is conditioned by a number of powerful forces. Vast media outlets owned by even more vast private corporations control and filter the mass media in defense of the status quo and their own interests. These corporations also sponsor many billions of dollars in advertising daily to condition potential consumers to buy or want to buy ever-more unsustainable goods and services. These goods and services themselves are manufactured or designed to be used up, to break down, or to be thrown away so that consumers will continue to buy new goods and services without end.

It is important to notice, here, that unsustainability is connected with structure. Our dominant institutions condition and insure unsustainable practices. It is not that their employees do not care about the future. Rather, the structure of the organizations that employ them predetermine a restricted range of thought and behavior. The organizations are structured so that their successful operation and survival are incompatible with sustainability.

Similarly, the industrial-military complex puts forth immense quantities of propaganda promoting a fear mentality and a corresponding security and war mentality. Huge numbers of scientists, engineers, and manufacturers around the world are employed in creating ever-more weapons of war, and ever-newer weapons of unimaginable destructive capability. This aspect of corporate capitalism promises the most lucrative investments, and nearly every multinational corporation is involved in the industrial military complex in one way or another.

At the same time nation-states around the world encourage loyalty to the nation and pride in being born to this or that fragment of humanity. They all encourage "supporting our troops" with special respect and legitimacy accorded to the men and women who "serve" the nation in this noble way. While the world sinks into planetary environmental disaster, vast sums of the world's wealth are poured down the toilet of militarism annually. Again, the structural features of these institutions predetermine a certain range of results.

All the while, the interlocked world system of some 193 so-called "sovereign" nation-states and global corporate capitalism lies in the background of all surface events and activities—silent, unquestioned, and assumed—as if these social phenomena were natural features of the Earth or inevitable laws of nature. Human beings everywhere are conditioned by this system, not only psychologically, but their working lives and necessity to survive requires that they participate in this system and support it in one way or another. Back in the 19th century, Karl Marx pointed out that unless human beings become conscious of their situation, they will be forever trapped within that situation. Unless they see what the system has done and is doing to them, they are fated to doom by the system itself. That is also our situation today.

This book shows the deep connections between our collapsing global ecosystem and our current world system of militarized nation-states and globalized corporate capitalism. It also explores the necessary connections between an ecologically sustainable planetary civilization and the absolute need for economic and political structures of our world social system that will make this possible. It makes very clear that we cannot deal with our lethal environmental crises without simultaneously dealing with the vast systems that underlie and cause those crises: global corporate capitalism and the system of militarized sovereign nation-states. And it shows in detail what the necessary social contract for a transformed world system would be like, how it would work, and the practical steps we can take to make it happen. This book, therefore, lays out the groundwork for a sustainable planetary civilization—our last hope, and our only hope.

This volume makes clear that our situation is not at all hopeless. Human beings are in the midst of a paradigm shift from a world view that is inherently fragmented and mechanistic to a new paradigm that is inherently holistic, ecological, and premised on unity in diversity. This new paradigm is examined and elucidated at length in order to make very clear what it entails and how it bears on the creation of a sustainable planetary civilization. But a new paradigm cannot involve only psychological and cultural changes on the part of human beings. It must also involve real structural change. For just as certain institutions are structured so as to predetermine unsustainable results, so properly designed institutions will predetermine sustainable results.

We examine below how this emerging holistic paradigm is embodied within the *Constitution for the Federation of Earth*, a document that can serve as a blueprint for a transformed world system—democratic, peaceful, prosperous, and sustainable. This book shows the necessary connections between a quality world economic system, a democratic governmental system, and the possibility of an ecologically sustainable civilization. It also shows the practical steps that we can take to make it happen.

Chapter 1

Our Present Danger and its Solution

1.1 Our Collapsing Biosphere Endangers Everything

THIS book, therefore, is about ecological sustainability in relationship to global economic and governmental systems. As such, it will not discuss our endangered planet or imperiled human future at great length. The earlier volumes in the series *The Sustainable (Development) Future of Mankind* by Dr. Timi Ecimovic and others has already accomplished this task, in addition to the many warning voices that have raised the alarm for more than half a century since the appearance of Rachel Carson's book *Silent Spring* in 1962. In this initial section, therefore, I will briefly remind the reader of the gravity of our situation before moving on to explore the necessary steps that we must take to transform that situation to a new sustainable world system.

James E. Hansen, the internationally known climate scientist who directs the NASA Goddard Institute for Space Studies and first modeled climate change on the supercomputers of NASA in

1

the 1980s, raised a dire warning in his testimony before the U.S.
Senate in 1988. In a Washington Post article of August 2012, he
writes:

> When I testified before the Senate in the hot summer of
> 1988, I warned of the kind of future that climate change
> would bring to us and our planet. I painted a grim pic-
> ture of the consequences of steadily increasing tempera-
> tures, driven by mankind's use of fossil fuels. But I have
> a confession to make: I was too optimistic. My projections
> about increasing global temperature have been proved true.
> But I failed to fully explore how quickly that average rise
> would drive an increase in extreme weather. In a new anal-
> ysis of the past six decades of global temperatures, which
> will be published Monday, my colleagues and I have re-
> vealed a stunning increase in the frequency of extremely hot
> summers, with deeply troubling ramifications for not only
> our future but also for our present.
>
> (http://www.washingtonpost.com/opinions/climate-
> change-is-here–and-worse-than-we-
> thought/2012/08/03/6ae604c2-dd90-11e1-8e43-
> 4a3c4375504a_story.html?hpid=z3)

As I write this in January 2013, 300 U.S. scientists, from both
the private and governmental sectors, have issued a 1000 page report on the present and pending effects of climate crisis on the
U.S. (Goldenberg 2013). The future will be significantly hotter,
drier, more disaster prone and may lead to epidemics, social unrest, shortages and disastrous storms throughout the U.S. The scientific community represents a compelling consensus of experts.
The voices of the experts have been warning us for the past half
century, yet little action has been taken.

Biospheric collapse is having (and will have) immense consequences for humanity. It is dramatically increasing the present
vast shortages in agricultural land, clean water, and other essential resources at the same time that the world's population continues to explode. It is causing great social unrest in many places
around the world (with increases in terrorism, rebellions, and

rogue armies) due to the scarcity of basic resources necessary to life. It will soon displace hundreds of millions of people who live in the present coastal areas of the planet. It is playing havoc with insurance systems and social systems as ever-more super-storms destroy areas around the world. It is causing tremendous die-offs in the biodiversity that is essential to the planetary food chain and ecological balance. It will destroy democracy and evoke greater militarization of the planet as nations strive to put down rebellions and keep order amidst the growing chaos.

In his 1989 book, *Entropy: Into the Greenhouse World*, Jeremy Rifkin describes the compelling need to shift global society from the false assumptions of the "Industrial Age" in which the economy is assumed to operate in a sphere independently of the ecosystem of our planet. The Industrial Age assumed that production and consumption could function in an endless "growth" mode independent of our planetary ecosystem that is governed by the law of entropy. He writes:

> The industrial age has been characterized by an exponential flow of energy and matter through the economic system. We have been extracting, processing, and discarding energy and matter faster than the earth's ecosystems can recycle the waste and replenish the resources. The buildup of polluted waste in the form of dissipated energy and organic and inorganic garbage now threaten the survivability of the earth. The statistics are grim. (1989: 192)

Nations still measure economic success in terms of growth or GDP. Businesses still measure success in terms of profits made without subtracting the negative social and environmental costs of generating that profit. We have created this vast industrial machine and corresponding ever-expanding economic machine that are now understood to be purely self-destructive and in contradiction with our finite, interdependent ecosystem. "Now our world and social system are falling victim to the very process of their creation," Rifkin writes. "Everywhere we look, the entropy of our world is reaching staggering proportions. We have become

creatures struggling to maintain ourselves in the midst of growing chaos. Each day we experience the truth that biologists have
long known: an organism cannot long survive immersed in its
own waste" (339-40).

"The statistics are grim," Rifkin declares. In her 1992 book,
If You Love This Planet: A Plan to Heal the Earth, Helen Caldicott
presents many of these grim statistics chapter by chapter: ozone
depletion, greenhouse effect, atmospheric degradation, disappearing forests, toxic pollution, species extinction, and overpop-
ulation. She then exposes the global economic system of "first
world greed and third world debt" and how the system of polluting and destructive consumerism is kept in place by the mass
media system of pervasive advertising and systematic disguise of
the real consequences of this system. She writes:

> You see, in terms of the biology of the planet development
> is a euphemism for destruction. Even the frequently used
> term sustainable development involves an exercise in con-
> fusion. In a world where all resources are finite—forests,
> minerals, soil, air, and water—continued use and abuse of
> them can only have one end: the depletion and destruction
> of most life. (1992: 43)

In his 2004 book, *Red Sky at Morning: America and the Crisis
of the Global Environment*, James Gustave Speth presents a very
scholarly and through account of the multiple dimensions of the
global environmental crisis. Speth is dean and professor at the
school of Forestry & Environmental Studies at Yale University and
was a scientific advisor on environmental issues for US presidents
Carter and Clinton. He is not, in any sense of the world, a radical
with an ax to grind. Speth presents the scientifically indisputable
grim statistics for our "ten global-scale concerns": ozone layer depletion, climate change, desertification (agricultural and grazing
lands worldwide turning into deserts), deforestation, biodiversity
loss, population growth, diminishing freshwater resources, ma-
rine environment deterioration (dying or dead fisheries in every

part of the world), toxification (tens of thousands of dangerous artificial chemical compounds in every environment on Earth), and acid rain.

These ten forms of global destruction, Speth says, are interactive and reinforce one another, portending immense disasters for the Earth. "Climate change," he writes, "could spark enormous transformations of uncertain probability that would be devastating to human societies" (2004: 210). He writes:

> A quarter century ago, scientists and others sounded the alarm regarding a set of linked threats to the global environment. Governments were put on notice, and, indeed, they acknowledged the issues and the need to address them. Yet the rates of environmental deterioration that stirred the international community continue essentially unabated today. The disturbing trends persist, and the problems have become deeper and truly urgent. The steps that governments took over the past two decades represent the first attempt at global environmental governance. It is an experiment that has largely failed. (2004:1-2)

Internationally recognized biologist John Cairns, Jr., maintains a website which he continually updates with the latest "grim statistics" and his scientific understanding of their implications. Like Speth, he writes about the US governmental responses to the facts and alarms raised by the climate scientists, which are not only inadequate but devastating for the future and its prospects:

> World-class climate scientist James Hansen has warned Congress that the planet has long passed the "dangerous level" for atmospheric greenhouse gases. Twenty years ago, he warned Congress about the consequences of "business as usual," and his predictions have been validated by the Intergovernmental Panel on Climate Change (IPCC), the national Academy of Sciences, and individual climate scientists. The Arctic ice melted 100 years ahead of IPCC predictions. Hansen affirms that a frank assessment of scientific data provides a certainty on climate change exceeding 99 percent.

Mark Lynas' book, *Six Degrees*, provides examples of what will happen for each degree of Celsius temperature increase up to 6 degrees (10.6 degrees Fahrenheit)—for example, a climate that is only a degree or so warmer than today could melt enough Greenland ice to drown coastal cities around the planet.

Politicians have set dates, such as 2025 and 2050, for reducing greenhouse gases, but Mother Nature neither bargains nor forgets transgressions. Violate her universal laws and penalties are usually severe and immediate, including starvation, disease, and death. Each day of delay in conforming to Mother Nature's laws forecloses options for providing a quality life for posterity and drives more species that exist as fellow passengers on Spaceship Earth to extinction. (http://www.johncairns.net/Commentaries/VT Research Intro.pdf)

Cairns writes that "the catastrophic effects of 'business as usual' have already become apparent." He presents the scientific facts under the following headings: the increasingly probable collapse of the present biosphere, the increase in catastrophic storms, droughts, and floods, the increased probability of pandemic diseases, the diminished productivity of renewable resources, including food, the catastrophic release of hazardous materials (such as the Fukushima disaster), the destruction of humanity's infrastructure, such as transportation, power, and food delivery systems, increased climate variability (such as temperature) and episodic storm release of hazardous materials. He writes that "anthropogenic greenhouse gas emissions continue to rise; and biodiversity loss and biotic impoverishment continue, as does exponential human population growth, oceanic acidity that may reach corrosive levels in the polar regions, and disparity in wealth. If the present Biosphere collapses, even the wealthiest one percent of the population will have no defense against the consequences." (http://www.johncairns.net/ebook2/Chapter40.pdf)

Similarly, global thinker B. Sidney Smith writes:

It is not a question of *if* we will overshoot our environment, and it is not a question of *when*. We are in overshoot—right now. Sometime in the 1970s came the first year when the *ecological footprint* of humanity became so large that the impact on the planet was no longer sustainable. Now, some 30 to 40 years later, we use up about 150% of the planet's yearly supply of resources every year. (2012: 24)

With all these top scientists and thinkers presenting an identical assessment of our situation and raising an urgent alarm, there appears little room left for not taking them very seriously. Our endangered planet and endangered future require immediate planetary wide programs premised on both sustainability and restoration, as far as is still possible, of our seriously damaged biosphere. How this is possible, in an efficient and practical manner, is the subject matter of this volume. But before we examine the possibility of genuine sustainable governance for the Earth, let us discuss the concepts of holism, ecology, and sustainability in order to deeply comprehend their meaning and implications.

1.2   Paradigm Shift from Fragmentation to Holism

Holism comprises the most fundamental understanding of our world order emerging from the 20th-century revolutions in science. This holism can transform our entire way of thinking and living on the Earth. It can become the basis for a new economics, a new ethics, and a new understanding of human social and political life. Yet these new understandings simply represent the fulfillment of certain civilizational fundamentals—like democracy—that go back to the ancient world. This paradigm shift in human thinking has not yet taken root in our ethical, social, or institutional life. We remain trapped in the older paradigms predicated on fragmentation and division. Our immense suicidal problems of the twenty-first century stem from this fragmentation.

Universally—in quantum theory, cosmology, ecology, systems theory, social science, and psychology—part and whole have come to be understood as inseparable from one another. The very meaning, structure, and function of the parts have become incomprehensible apart from the wholes within wholes (fields within fields) within which the parts are embedded and in terms of which their nature, evolution, and functioning must be understood. Yet our thinking remains mired in divisions, separations, and fragments that appear incommensurable with one another. The result is collective and personal egoism, war, conflict, economic exploitation, destruction of nature, and destruction of one another.

Science has revealed that, at every level, distinct entities, individuals, are part of an interrelated matrix of organized matter and energy societies: fields that relate the individuals to one another in a multiplicity of ways and distinguish them as distinct individuals embedded within the fields. In other words, individuals are not only contradictory to one another in the sense that *A* and *not-A* appear as logically mutually exclusive. *They are simultaneously complimentary to one another* as instances of a more encompassing set of universals or fields (Harris 2000a; Martin 2008: Ch. 3).

Holism means that we must enlarge our thinking to encompass the manifold of fields within which we are embedded. No longer is *A* simply incommensurable with *not-A*. A clear view of reality requires that I discern the ways in which *not-A* is complementary to *A*. The Other person is inseparable from the very possibility of my existence, since it is the fields within which we are embedded that make possible the existence of the Other and my existence. The Other does not contradict me in an irreconcilable manner, but becomes complementary to me as another essential part within a more encompassing whole. The outmoded logic of sets is superseded by the dialectical logic of wholes (Harris 1987).

The other person, indeed, remains a center of moral freedom that cannot be reduced to any scientific or behavioristic set of compulsions or bio-chemical reactions. The absolute dignity of the Other derives from this fact, as will become clear below. However, other persons and I interpenetrate and overlap in a vast multiplicity of ways that unify us as human beings within our common moral and civilizational project. Today, we have also realized that our common civilization project includes the precious Earth on which we dwell—its beauty, its ecological integrity, its fragile biosphere, and its proper, holistic governance.

In his book *The Systems View of the World: A Holistic Vision for Our Time*, systems theorist Ervin Laszlo calls this structure of our world "holarchy." The holism of individuals flourishing within the fields that sustain and make possible their individual existence is reflected in a hierarchy of wholes within more encompassing wholes within still greater wholes from the sub-atomic level to the level of the cosmos. He writes:

> A holarchically (rather than hierarchically) integrated system is not a passive system, committed to the status quo. It is a dynamic and adaptive entity, reflecting in its own functioning the patterns of change over all levels of the system. The holistic vision of nature is one of harmony and dynamic balance. Progress is triggered from below without determination from above, and is thus both definite and open-ended. To be "with it" one must adapt, and that means moving along. There is freedom in choosing one's path of progress, yet this freedom is bounded by the limits of compatibility with the dynamic structure of the whole in which one finds oneself. (1996: 58)

Human beings are integral parts, not only of the holism of the cosmos and the ecosystem of the Earth, but of one human species and planetary society encompassing the Earth. However, in practice, just as we have not yet harmonized our civilization to the delicately balanced biosphere that sustains all life on our planet, so we have not harmonized our social life to the holism of planetary society. We remain trapped in systems of fragmentation that are destroying the biosphere and continue to destroy planetary society through war, patterns of exploitation, linguistic forms of deceit, organized violence, and perpetual conflict.

Fragmented systems and fragmented patterns of thought go hand in hand. The holistic view of the cosmos and human life emerging from 20th century science has not yet been assimilated into a paradigm shift of the ways that we think and organize our political, economic, and cultural lives. Laszlo says that holarchically organized systems are influenced from below, not determined from above. However, the influence is *reciprocal* in any truly holistic system, for in a true whole the parts function as integral elements in the functioning and maintenance of the system as a whole. The principle of wholeness structures the relationships among the parts, not diminishing individuality but rather actualizing a genuine individuality in which the complementary functioning of the parts integrates and maintains the integrity of the whole. In this section we will primarily look at the holism of humanity. In subsequent sections, we will examine holism in relation to the natural sciences.

When applied to human life and ethics, thinkers who understand the principle of holism advocate *linking our individual lives with all human beings*, since our humanity, our oneness with all other persons, is inseparable from our uniqueness as individuals. Our ability to link our lives in this way means discovering our own fundamental humanity. As we become ever-more fully human, we begin to realize that nothing and no one separates me from the others. Our fragmented sense of self *that defines itself in opposition to what it is not* begins to give way to a deeper sense of self that lives from the universality of its own humanity. The higher potential of our self is progressively actualized. This process of actualization of what is universal within ourselves is affirmed by many leading psychologists and thinkers.

Psychologist Erich Fromm affirms that this "means a constant striving to develop one's powers of life and reason to a point at which a new harmony with the world is attained; it means striving for humility, to see one's identity with all beings, and to give up the illusion of a separate, indestructible ego" (1962: 156). Psychologist Robert J. Lifton writes "One moves toward becoming what

the early Karl Marx called a 'species-being,' a fully human being. Once established, the species identification itself contributes
to centering and grounding. In no way eliminated, prior identifications are, rather, brought into new alignment within a more
inclusive sense of self" (1993: 231).

Similarly, psychologist and spiritual teacher Richard De Martino affirms that "to the degree to which I can rid myself of this
filter and can experience my self as the universal man, that is,
to the degree to which repressedness diminishes, I am in touch
with the deepest sources within myself, and that means with all
of humanity" (1960: 127). Spiritual teacher Jiddu Krishnamurti
declares that "if you don't know how your own mind works you
cannot actually understand what society is, because your mind is
part of society; it is society.... Your mind is humanity, and when
you perceive this, you will have immense compassion" (1989: 83-
86). For such spiritual teachers, careful attention to the workings
of our own consciousness and our common human situation inevitably illuminate for me my identity with all humanity.

Mahatma Gandhi was also a holistic thinker who understood
that each unique person is an expression of the whole, an expression of "Truth." His fundamental principle of ethics and nonviolence was *satyagraha*, literally "clinging to truth." If we respect
the uniqueness of each, instead of privileging our own differences that set us apart from the others, then the truth of the whole
will begin to emerge. Our unity, our mutual participation within
larger wholes, will begin to become clear to all concerned. We
shall examine below some fundamental links between democracy,
nonviolence, and holism.

In the 18th century, Immanuel Kant affirmed that "rational beings all stand under the law that each of them should treat himself
and all others, never merely as a means, but always at the same
time as an end in himself.... morality consists in the relation of
all action to the making of laws whereby alone a kingdom of ends
is possible" (1964: 100-101). The kingdom of ends as a command
of morality, for Kant, means that each of us adopts moral law for

ourselves with a view to a world in which every person treats every other as a unique and infinitely valuable "end in himself," a world under universal moral laws in which everyone treats everyone else with unreserved respect and dignity. Individual moral reasoning inevitably links us with all others. Morality directly connects us with the holism of humanity. Following Kant, G.W.F. Hegel, developed this holism by embodying it concretely within the whole of society, showing the interrelation of part and whole at every level of society.

Out of the dozens of thinkers affirming ethical and human holism since the 20th century, I will cite just two more. Throughout his long lifetime of philosophical output, John Dewey affirmed the inseparability of the individual and the community, ultimately the human community as a whole, as the matrix for freedom and the development of our individual potential. For Dewey, the concept of democracy itself simply "projects to their logical and practical limit forces inherent in human nature" (1963: 497). The democratic ideal is simply a projection of our common human potential beyond, for example, "the secondary and provisional character of national sovereignty" (1993:120). Dewey's life work articulates the holism of humanity and its common ideal of an ever-greater actualization of our potential for free and open association with one another within the matrix of our planetary community.

Similarly, philosopher Errol E. Harris affirms that "in human self-awareness, the *nisus* to the whole has become conscious of itself, so the self, being apprised of its own desires and their aims, strives to organize them, in order to attain coherent wholeness, in which it can find complete self-satisfaction; that is, to make them mutually compatible, so as to remove the frustration inherent in internal conflict. It is this self-realization that determines the ultimate standard of value" (2000b: 251). The universal drive at the heart of the evolutionary process (its *nisus*), Harris states, operates in us (as it does everywhere) to promote wholeness, holism: the removal of internal and external conflict so that the individual person (or group or nation) and the human species can live at peace within a dynamic and diverse yet ordered whole. Its standard is *reason*, a reason discerning the holistic character of the world process and progressively conforming our lives and institutions in harmony with it. However, its dynamic includes an integration and harmonization of the *whole person*: thoughts, emotions, intuitions, customs, habits, and instincts.

Laszlo argues that because human beings are self-aware, goal-oriented creatures, all of our ends or purposes constitute value-oriented activity. Our highest value involves the fulfillment of our potential as individual human beings, to become what we are capable of being, which is not possible without the matrix of society and civilization of which we are inseparable parts. Holistic values, therefore, seek to actualize self-fulfillment within the empowering framework of the larger social wholes that encompass us and make our self-actualization possible. It follows that fragmented economic, political, and cultural institutions defeat or interfere with this process. All the above cited thinkers conclude that we must link our individual lives with universal humanity. Our individuality and our humanity become inseparable.

This insight is absolutely important because the ethical and social holism affirmed by these thinkers mirrors the ecological holism revealed by science and thinkers concerned with biospheric sustainability. The volume *Valuing the Earth: Economics, Ecology, Ethics* (1993), edited by Herman E. Daly and Kenneth N. Townsend, brings together a collection of writings by both scientists and ethical humanists revealing the deep parallelism between human and ecosystem holism. Fundamental to both is the fact that relations are *internal* rather than *external*. The notion that relations are primarily *external* derives from the early modern paradigm in which the world was thought to be composed of "substances" existing independently of other substances. This was primarily an atomistic model in which everything was reducible to the substantial, independent atoms of which things were composed.

The 20th century revolution in all the sciences gave us an entirely different model, one in which the parts can only be understood in terms of the "fields" or wholes of which they are part. Relations consequently become internal relations. The part could not exist, nor be what it is, apart from its place and set of relationships within the whole. In *The Liberation of Life* (1990), Charles Birch and John B. Cobb, Jr. write:

> The ecological model proposes that on closer examination the constituent elements of the structure at each level operate in patterns of interconnectedness which are not mechanical. Each element behaves as it does because of the relations it has to other elements in the whole, and these relations are not well understood in terms of the laws of mechanics. The true character of these relations is discussed in the following section as 'internal' relations. Internal relations characterize events. For example, field theory in physics shows that the events which make up the field have their existence only as parts of the field. These events cannot exist apart from the field. They are internally related to one another. (1990: 83 & 88)

What kind of institutions would reflect this holism and these internal relationships? The most basic answer is familiar yet strange to us: *democracy*, properly understood. The uniting of persons within a democratic community in which people understand that they are bound to one another by internal relationships transitionally known as their social contract. Democracy, properly understood, is the foundation-stone for human beings coming together to establish a sustainable civilization in harmony with the holistic principles of ecology. In the face of the immense terrors of our time, and the on-going collapse of our planetary ecology, we need to understand the fragmentation of our thought and our outdated institutions. We must act to discover the holism within ourselves and how it might be reflected in holistic, nonviolent, and sustainable institutions. My contention is that uniting humanity under a *Constitution for the Federation of Earth* constitutes our best hope for creating a human holism capable of establishing a civilization in harmony with our planet's ecological holism.

Just as the fish living in a coral reef off one of the seacoasts of the world exist within a set of internal relationships to the other fish and their environment—to water temperature, to the health of the reef, to the rhythm of the tides, the seasons, the protection of the ozone layer, etc.—so human beings exist within a set of internal relationships both to their natural environment and to one another and human civilization. Thinkers such as philosopher Jürgen Habermas (1998) and language scientist Steven Pinker (1995) have shown the universality of language and its necessary connections with the selfhood of each of us and the communities that sustain and make possible our individual survival and flourishing. Yet nation-states are structured as if their relations to the rest of the world were external, and corporate capitalist competition and exploitation (of people and the environment) are structured as if the other corporations, their exploited victims, and the natural world were external to their success and autonomy.

The current, horrific global economic crisis is a consequence of fragmentation, of a lack of holism and democracy in our institutions. The patchwork attempts at solutions by the Obama administration and other national actors worldwide will inevitably result in failure. The current disastrous world order of poverty, misery, war, and violence is likewise a consequence of lack of holism and democracy, of thinking in terms of external rather than internal relationships. The disintegrating integrity of our environment is only a reflection of a global economic and political disorder premised on an outdated mechanism and atomism deriving from the early modern paradigm, now hopelessly outdated. Our survival on this planet, along with the future of our children and other precious living creatures, depends on our ability to establish holistic institutions and holistic patterns of thought within the very near future.

We shall see that these insights form the foundation for rapidly moving to authentic planetary democracy, beyond the dogma of "sovereign nation-states" that see themselves as independent of the rest of humanity. If we understand the holism affirmed by every 20th century science, and we simultaneously understand the fragmentation of our present modes of thinking and our cultural, political and economic institutions, we will comprehend the absolute imperative to establish planetary democracy as efficiently as possible. It is planetary democracy that provides the holistic framework for each human being to realize his or her potential to the maximum extent possible—by guaranteeing equal rights and equal freedom to everyone with a matrix of common social, political, and economic institutions. Planetary democracy embodies the holism that is necessary in three essential ways: for our ecological survival on this planet, for the progress and fulfillment of the historical human project, and for our personal fulfillment as individuals.

In a manner similar to all natural systems, planetary democracy will function as a "holarchy." Local communities interact democratically and economically, addressing local problems and issues within a federated world order. Larger regional social and political units (for, example, cantons of China, pradesh of India, or states within the United States) also function democratically, dealing with regional problems and issues. Nations "holarchically" include these smaller units and are themselves included within the Earth Federation that addresses planetary problems and issues, again through democratic processes: through protection of the rights of individuals and the federated units within the system and through enforceable laws maximizing the equal freedom of each to develop his or her potential within a framework of the common good of the whole of humanity and our planetary ecosystem.

Another aspect of genuine wholes is that the structure of the whole provides a *telos* (goal) for the operation of the parts individually and collectively. The *telos* of the heart, liver, and lungs in the human body is the proper functioning of the whole body. The heart, liver and lungs cannot be understood apart from their roles within this *telos*. And their functioning as individual parts is evaluated in terms of the degree to which it contributes to this holistic process. In a world governed by a democratic constitution that integrates all nations, persons, and communities into its functioning, the purposes of the whole will influence the functioning of the parts. A constitution premised on creating a sustainable planetary civilization will influence the operation of the parts accordingly. The present chaos of competition, greed, fear, and self-interest will be transformed by this new paradigm. Holism will replace fragmentation.

This describes in a very brief way the coming great transformation: the rebirth of human civilization that will either develop rapidly, signaling the happy survival and flourishing of the human historical project, or will happen not at all because the human project will have ended in major planetary disaster or possible extinction of our species. We will see below a number of ways in which democracy has been *misunderstood* as well as a number of ways that it has been *manipulated* to prevent its genuinely holistic potential from emerging. We also will come to understand that violence against persons and the environment *diminishes* in our world to the extent that democracy is realized. We will see that democracy, sustainable economics, and environmental concern go hand and hand in any effective form of governance. Their actualization on Earth will mean the beginning of a sustainable planetary civilization.

1.3 Ecology and the Science of Sustainability

Definitions of sustainability vary in emphasis and focus but nearly all share a basic framework such as the following: "Meeting the needs of the present without destroying the ecosystem of the Earth and depleting its resources so that future generations are unable to meet their needs," or, more simply, "the ability to flourish in the present without diminishing the ability of future generations to flourish." We can only meet the needs of the present

without destroying the life-prospects of future generations by recognizing our interdependence with the planetary biosphere. The
concept of sustainability, therefore, participates in the holism that
we have seen as the defining feature of the 20th century revolutions in science. Ecology as a science focuses on the necessary
interdependence of organisms and their environment.

At the very least the concept of holism requires a new economics for the people of Earth, something that is being worked
on by many non-mainstream economists such as Lester R. Brown,
Hazel Henderson, David Korten, Herman E. Daly, and Kenneth
N. Townsend. In their 1993 book *Environmental Economics*, Turner,
Pearce, and Bateman affirm that "conventional economics text-
books often convey a very misleading picture of the relationship
between an economic system... and the environment.... Basically,
simple economic models have ignored the economy-environment
interrelationships altogether" (15). Sustainability means precisely
that the economy cannot create goods and services to support human life without understanding that the economy is a subset of
the biosphere of the planet. To date, economics has been prac-
ticed as if investment, production, and consumption were self-
contained, without thought of the immense negative external consequences that production and consumption have on society and
the planetary ecosystem.

The concept of sustainability refers to the ability of human
beings to integrate their societies and economies into the larger
holism of the Earth's biosphere and ecological balance, which
includes an ethical and social holism linking them to the well-
being of future generations. Economics will have to change, to
be sure, and this will have to include transformed patterns of investment, production, engineering, design, transportation, distribution, consumption, waste production and disposal, etc. It will
require changes in education, social media, and the way we think
about life and one another.

Holism will mean transformed civil institutions, organizations, ways of working together, communicating and doing busi-

ness. It will necessarily mean the ending of militarism, which, in all its phases, from production to deployment to use of military organization and weaponry, is utterly unsustainable and wasteful. Politics will have to change as well, from a politics of wealth, power, and winners-take-all to a politics of cooperation, fairness, and inclusiveness, in other words, to authentic economic and political democracy. And democracy will have to be globalized. It will necessarily have to be planetary democracy.

Non-renewable resources will be used at a bare minimum. All energy will have to be clean and renewable (sun, wind, tides, geothermal, etc.). Everything will have to be manufactured for maximum durability and reparability. Waste (thermal and trash) will be reduced to a minimum and whatever waste there is will have to be biodegradable and non-toxic. What little toxic wastes still produced will have to be detoxified. Forests will have to be re-planted, watersheds repaired, and biodiversity protected. World-wide agricultural lands will be cultivated with minimum soil loss and programs of soil restoration, while using natural insect repellants and natural, biodegradable fertilizers.

The implications are vast, for they require a transformation of human civilization throughout the entire range of our practices. In their book *The New Science of Sustainability: Building a Foundation for Great Change*, Goerner, Dyck, and Lagerroos elaborate on this point: "Sustainability began as a subset of environmentalism, as concern about how to manage the flow of natural resource inputs and human-generated outputs (pollutants and waste) in a way that could go on, if not forever, at least for a very long time. Yet, when people began to think about how to alter modern practices, they began to realize that the changes needed would require a major shift in the way we *think* about nature, agriculture, energy, economics, health, and, from there, ethics, democracy, and social justice" (2008:31).

We can see, therefore, that economist Herman E. Daly's definition of sustainability as "development without growth beyond environmental carrying capacity, where development means qualitative improvement and growth means quantitative increase" (1996: 9) carries with it vast ramifications—for education, economics, culture, and government. In this book, called *Beyond Growth*, Daly demonstrates that the perpetual growth model of mainstream modern economics systematically ignores the carrying capacity, finite resources, and fragile ecosystem of the Earth in its economic formulas. As Jeremy Rifkin puts this: "making the transition from a nonrenewable to a renewable base of energy represents a monumental task for the whole of civilization" (1989: 210)

Daly understands the pervasive holism of the new scientific paradigm and he applies this directly to sustainability or what he calls "the steady-state paradigm."

> Ecology is whole. It brings together the broken, analyzed, alienated, fragmented pieces of man's image of the world. Ecology is also a fad, but when the fad passes, the movement toward wholeness must continue. Unless the physical, the social, and the moral dimensions of our knowledge are integrated in a unified paradigm offering a vision of wholeness, no solutions to our problems are likely. (1996: 357)

In his 1996 book, *The Web of Life: A New Scientific Understanding of Living Systems*, Fritjof Capra characterizes the ecological model in terms of five principles: interdependence, cyclical feedback loops (recycling), pervasive cooperation, flexibility, and diversity (298-304). All these concepts are interconnected and linked with one another. Interdependence, therefore, can be said to imply all the others. Capra writes:

> Interdependence—the mutual dependence of all life processes on one another—is the nature of all ecological relationships. The behavior of every living member of the ecosystem depends on the behavior of many others. The success of the whole community depends on the success of its individual members, while the success of each member

depends on the success of the community as a whole. (1996: 298)

Can human beings begin thinking in terms of cooperation and interdependence rather than competition, hostility, fragmentation, and militarized defense? The challenge demands that we make a complete and genuine paradigm-shift from our current divisions and fragmentation to holism. In his 2001 book *Eco-Economy*, Lester R. Brown calls the change that is needed "Copernican," as great as the shift from the geocentric to the heliocentric world views (21). Such a fundamental shift in our world-view is clearly an absolute necessity if there is to be a viable future at all.

In *Only One Earth*, Barbara Ward and Rene Dubos state that the transformation of human consciousness is indeed possible, affirming "that men can experience such transformations is not in doubt. From family to clan, from clan to nation, from nation to federation—such enlargements of allegiance have occurred without wiping out the earlier loves. Today, in human society, we can perhaps hope to survive in all our prized diversity provided we can achieve an ultimate loyalty to our single, beautiful, and vulnerable planet Earth" (1972:220). But the issue is perhaps deeper than this for it involves the way people's loyalties are shaped by dominant institutions. Brown raises this issue in the following way:

> The central question is whether the accelerating change that is an integral part of the modern landscape is beginning to exceed the capacity of our social institutions to cope with change. Change is particularly difficult for institutions dealing with international or global issues that require a concerted, cooperative effort by many countries with contrasting cultures if they are to succeed. For example, sustaining the existing oceanic fish catch may be possible only if numerous agreements are reached among countries on the limits to fishing in individual oceanic fisheries. And can governments working together at the global level, move fast enough to stabilize climate before it disrupts economic progress? (2001:20-21)

Above we saw the statements of climate scientists concerning the failing ecosystem of our planet as of 2013. And we saw that the scientific evidence has been overwhelming and continuously mounting for at least the past 50 years. If the behavior of nations to date is any indication, the answer to Brown's questions is clearly "no". The system of sovereign nations will not be able to take the necessary measures to save the ecosystem of the Earth. Yet in his book Brown himself appears unable to imagine any alternative other than the immensely unlikely cooperation of some 193 militarized nation-states in addressing our multiplicity of global environmental crises.

The same is true of the environmental economics of Turner, Pearce, and Bateman. Their book, excellent as it may be in other respects, still treats sustainable economics as if it were a matter of adding up multiple, separate national accountings. They state that "the current system of national accounts used in many countries fails, in almost all cases, to treat natural capital as assets which play a vital part in providing a flow of continuous output/income over time.... Extended national accounts...are required in order to improve policy signals relating to SD" [sustainable development] (1993:56). Like Brown, they describe the economic accounting system that must treat the economy as a subset of the natural environment, simply assuming that this will need to apply independently to each of 193 sovereign nation-states.

We have seen, however, that there are two gigantic dominant global institutions that must be transformed if we are to adopt a truly holistic planetary orientation: the first is global corporate capitalism and the second is the global system of so-called "sovereign" nation-states. Both are interdependent, interrelated and inseparable from one another in the modern world system as many thinkers have pointed out. It is quite astonishing that many of the economists who claim to be articulating a complete theory of sustainability and who speak of the need for a major transformation of our fragmented thought patterns seem unable to think beyond the clearly fragmented and unworkable system of sovereign nation-states.

Many thinkers and scholars have described the development of this modern world system, its internal dynamics and processes, over these centuries. By "modernity,"of course, we mean that period of world history since the European Renaissance during which the institution of the "sovereign nation-state" and the economic institutions of global capitalism developed as the dominant world institutions. The accounts of historians (such as Arnold Toynbee), sociologists (such as Max Weber), economists (such as Thorstein Veblen or Karl Marx), and philosophers (such as Jürgen Habermas) have nearly universally revolved around the interrelation of these two dominant institutions of the modern world. We have seen that these institutions have been interdependent for centuries.

The dominant features of the system are quite clear to many serious thinkers and scholars. While there is active debate as to the internal periodization in the development of the world system, as well as over the cyclical features of the system (e.g., how often and of what severity must depressions and recessions recur: Chase-Dunn 1998), the interrelation of the dominant institutions of the modern world has been described at length in volume after volume (e.g., Parenti 1995, Smith 2005, Petras and Veltmeyer 2005, Klein 2007). Yet the work of these thinkers has not led to a general understanding among the mainstream population of today's world, nor for many climate scientists. These gigantic institutions are so pervasive that they tend to form the invisible background of life as if they were pervasive natural phenomena, like the air we breathe or the water we often take for granted, and their self-justifying propaganda is everywhere.

What is not controversial and forms part of the awareness of most educated people is the understanding that the sovereign nation-state evolved out of the medieval feudal political system just as modern capitalism evolved from feudal economic relations. These institutions have operated as an inseparable set of interrelated systems, inextricable and interdependent. Their destructive consequences have operated in tandem since the very beginning. As social scientist Christopher Chase-Dunn expresses this:

> The state and the interstate system are not separate from capitalism, but rather are the main institutional supports of capitalist production relations. The system of unequally powerful and competing nation-states is part of the competitive struggle of capitalism, and thus wars and geopolitics are a systematic part of capitalist dynamics, not exogenous forces. (1998: 61)

This means that the multiplicity of negative consequences for our world system originate in the integral systemic nexus of these two institutions that are inseparably linked and determinative of the social, political, economic, and environmental conditions in which planet Earth is today mired. Michael Hardt and Antonio Negri analyze the relation between global expansion of capitalism and nation-state imperialism. It was the need for new markets and new ways of investing surplus value under the capitalist need to expand or die, that let the ruling classes of the imperial nations into foreign imperial and/or postcolonial conflicts (2000: 221-239).

People are also aware that these macro-features of the modern world system have developed through a number of overlapping and interrelated periods often identified by thinkers under such headings as the era of the conquistadors (or so-called primitive accumulation), the era of mercantilism, the era of colonization, the era of slavery, the era of decolonization and neocolonialism, etc. Are the immense crises that today threaten to tear our world asunder due to the largely unspoken and unexamined premises under which we have operated for the past several centuries, today hopelessly outdated and inadequate for our needs?

Much quality work has been done by thinkers of the 19th and 20th centuries attempting to extrapolate the observable and probable consequences of this system. The latter point is fundamental. Every set of political, economic and social assumptions carries within itself probable consequences. Some of the sections in this volume identify and elaborate these consequences, consequences identical with the factual conditions we find all about us in today's world. The *internal logic of the world system* results in certain political, economic, and social consequences such as commodification of existence, wide-spread poverty, perpetual wars, militarism, human rights violations, and environmental destruction.

Chris Williams, for example, in his 2010 book, *Ecology and Socialism: Solutions to Capitalist Economic Crisis*, affirms that "capitalism is thus systematically driven toward the ruination of the planet and we underestimate how committed the system is to planetary ecocide at our peril. As stated above, ecological devastation is just as intrinsic to the operation of capitalism as is the exploitation of the vast majority of humans in the interests of a tiny minority, imperialism, and war" (232). Yet he ignores the fact that capitalism cannot engage in imperialism and war alone, that it is its synthesis with the militarized imperial nation-states that causes these ravages, and that a fundamental change in capitalism will require a concomitant change in the system of sovereign nation-states.

Capitalism is inherently anti-ecological just as the nation-state is anti-ecological. The nexus of capital and the nation state means that the imperial centers of capital must expand, for the logic of capital is that it must perpetually expand and reinvest its surplus value, appropriating ever-more resources, institutions, and peoples into capitalist production relations and the creation of ever-more surplus value. This economic system cannot continue in its present form if human beings are to achieve a sustainable civilization for the Earth.

Ecologist Jeremy Rifkin correctly states that the concept of "private property" will have to change radically within a low-entropy, sustainable culture. While personal consumer goods and real estate, he says, will remain protected, large tracts of land and major resources will have to be public and directed toward the common good of all (1989: 245). But the public trust will have to include more than this. It will necessarily have to protect the oceans of the world, the atmosphere around the world, the ozone layer, and the vital natural resources of the world. What kind of global institutions are necessary if the global commons is to be protected on behalf of the people of Earth?

In his essay on "Environmental Ethics," Shridath Ramphal centers on a fundamentally correct evaluation of our situation. He writes that "today the concepts of neighbor and neighborliness are being enlarged and refined in our interdependent world, but have yet to be brought under the rule of law. What global interdependence means is that we all need each other in some measure: for prosperity, for subsistence, for survival. Our planet offers no sanctuaries. There are no shelters that insulate anyone, anywhere, from disease, from the effects of poverty, from military holocaust, from environmental collapse" (in Mayur 1996: 172). The rule of law does not have to empower capitalist economic relations of relentless expansion on a finite planet. It can structure an economics of sustainability, cooperation, and equity.

1.4   Positive Freedom in a Planetary Democracy

Ramphal correctly points out that neighborliness, interdependency, and global community will always remain incomplete until they are brought "under the rule of law." For all their talk about a paradigm shift in human consciousness, most of the thinkers cited above who discuss sustainability ignore the fact that there is no planetary democratic mechanism for planetary decision-making. Today, global decisions are made by huge multinational corporations and banks having no transparency and no accountability to the people of Earth. They are also made by powerful militarized imperial nations in terms of perceived "national self-interest," papered over by voluntary, unenforceable U.N. treaties. The U.N. is in fact premised on the "sovereignty" of its member states and is colonized by these same gigantic non-democratic global institutions: multinational corporations and imperial nation-states.

As many contemporary thinkers of the new paradigm have pointed out, we are today interdependent with all our neighbors in the global community, and our freedoms, as well as our futures, depend on our relationship with them. Yet they ignore that fact that democracy is the only form of decision-making that legitimately allows people to act as a whole, to make decisions for a common future and a common good. Unless our interdependent' relationship is governed by the rule of democratically legislated laws, how are we going to make decisions and coordinate globally to avert climate collapse? All our resources must be devoted to restoring and protecting the environment. Only the democratic representatives of the community of the Earth itself would be able to effectively and legitimately make the decisions necessary to protect our common future.

At present mere voluntary treaties among militarized nation-states, or supposedly voluntary cooperation among gigantic multinational corporations institutionally organized to promote private profit, or moralized lobbying from a plethora of non-governmental organizations—are all hopelessly inadequate forms of action in the face of our global climate crisis. Even if decisions and action are funneled through the U.N. system, they still involve merely a contingent coalescence of nations, corporations, and NGOs, and hence still remain an undemocratic and haphazard mode of decision-making and action entirely inadequate to our need.

There are a number of key international documents concerned to promote sustainability for the planet. These include *Agenda 21* (that emerged from the 1992 Rio de Janeiro U.N. Conference on the Environment), the Kyoto Protocol (adopted in 1997 and voluntarily signed to date by some 191 nations), and the Millennium Development Goals (eight goals, including environmental sustainability, adopted by all 193 U.N. member nations in the year 2000, who agreed to achieve these goals by 2015). Most of the nations signatory to these documents remain very far from achieving the goals they agreed to. The largest polluter in the world, the United States, never even signed the Kyoto Protocol. The fact is that there is simply no way to achieve sustainability under the current global economic system intersecting with a system of some 193 semi-autonomous nation-state entities.

How are all these militarized nation-states going to abandon their incredibly wasteful drain of their collective resources into militarism, or global debt service to private multinational banks, and simultaneously devote their resources to a conversion to sustainability? How are global corporations, devoted to systematically accumulating private profit for the wealthy 10% of the human population who hold the lion's share of investments, going to transform themselves into environmentally beneficial organizations devoted to the common good of the Earth?

The answer, of course, that a sustainable world system is impossible unless both the world's dominant institutions are converted to holism and interdependency. Both capitalism and sovereign nation-states must be transformed holistically. The only practical answer is one democratic world system that protects human rights and freedoms equally everywhere and converts the world's economic life to a market form of economic democracy directed toward universal equity, prosperity, and sustainability. Unless we achieve a sense of global community, where nearly everyone understands sustainability and cooperates with others to achieve it in every locality, climate collapse cannot be averted.

Anyone who has traveled widely in the world knows that human beings everywhere share a common consciousness. Despite tremendous variety in customs, languages, religions, and traditions, people everywhere understand one another on many levels. They are structured the same, function the same, have the same basic needs and wants, and in many ways think the same. Precisely because it is so pervasive, people often live unaware of this common consciousness or sameness.

Every person comes from the social unit of a family and has grown up as part of a larger series of communities from the neighborhood to the town to the region to the nation. Everyone's selfhood has emerged through these interactions and is a result of interaction between the person's individual biological and personal characteristics and the environing communities. Every non-impaired person speaks a language or languages. Every person is aware of themselves as a human being who lives as a resident and citizen of the Earth. These universal human characteristics tell us that a global community of cooperation and compassion is indeed possible.

Over the past several centuries philosophers have developed a profound alternative tradition to the negative freedom of philosophical liberalism, a tradition that understands freedom as a positive quality of human life that arises out of, and cannot be detached from, it roots in the communities in which people participate. This alternative tradition runs through such thinkers as Baruch Spinoza, Jean-Jacques Rousseau, Immanuel Kant, G. W. F. Hegel, T.H. Green, Bernard Bosanquet, Ernest Barker, Errol E. Harris, and Jürgen Habermas. Habermas has demonstrated through his analysis of language (perhaps our most fundamental social and communal capacity) the empirical fact that the individual selfhood of each person arises from the matrix of community and the universal procedural principles of justice, equality, and reciprocity that are necessarily presupposed by every user of language (1998).

Nevertheless, by showing that the basic procedures of democracy are integral to the presuppositions of the very possibility of language and human communication (and hence scientifically grounding democratic theory) Habermas is not adding substantially to the conception of democratic liberty developed by the other thinkers mentioned here. The presuppositions making language possible—the equal right to speak, the equal right of ideas to be considered, the equality of participants in the act of speaking, etc.—were already understood as the heart of democracy within this tradition. Ernest Barker (1874-1960), for example, writes concerning democracy:

> Men naturally concentrate their attention on these matters of regulation and the problems which they involve—problems such as proportional representation, the right of

dissolution, the right of referendum, and whatever else can be made a matter of formal legal right. But when constitutional law has done its utmost, it leaves a sphere which needs control, and yet cannot be controlled by legal rule. Discussion, by its very nature and in its own essence, transcends the scope of legal control. What it cannot transcend is the rules of its own inner logic and its own inward ethics—or rather it can only transcend them at the cost of annihilating itself. Discussion which refuses any control becomes a civil war; and civil war is the end of discussion. (1967: 72)

We are bound together on planet Earth in a community of discussion, debate, and collective decision-making. Just as language itself contains its "own inner logic and its own inward ethics," so genuine democracy has long been understood to embody these requirements. Our individual freedoms arise from communities that internalize and support these fundamental and universal human ethical conditions. The only alternative to discussion and cooperative decision-making is some form of coercion or strategic manipulation, necessarily involving violence or the threat of violence. Yet the Earth to date has no institutions for allowing our planetary community of discussion, debate, and decision-making to take action on behalf of the planetary common good.

Liberty arises from the common consciousness and supporting matrix of a community. If other persons are not in mutual support and do not make the kind of assumptions that support one another's liberties: if others resist or mistrust or limit relationships to impersonal "contracts," rigorous "security" measures, or atomistic economic interests, then liberty becomes seriously impeded. It becomes a perpetual fight to do what one wishes over and against the community, the government, the neighbors, or the law. Unless the community empowers my freedom through its internal relationships and basic assumptions, I am trapped in relationships of social and legal coercion against which I must struggle.

On the other hand, relations of mutual trust and understanding open up dimensions of liberty far beyond anything possible through the struggle of an individual over and against the community. Just as the essentially social nature of the self makes of the Lockean self (that assumes *a priori* rights and autonomy before giving its "consent" to civil society) a profound falsification of reality, so the negative idea that freedom can only be won through limiting government, resisting community, and distrusting neighbors is a profound falsification of freedom.

Positive freedom emerging within authentic communities is found within the context of a "general will" or mutually understood common good that affirms a diversity of individuals and their freedom precisely because these are rooted in the unity of the community supporting common assumptions about the rights and freedoms of each. Philosopher Alan Gewirth (1996) calls this the "community of rights." The French revolutionary slogan "liberty, equality, and fraternity" had it correct, since there is no liberty without both substantial equality and community. The liberal tradition takes only the first of these seriously, treating the second as merely an equality of rights for atomized individuals to be let alone by the state or other persons irrespective of vast economic inequalities, and largely discounting the third.

The very *concept of government* changes with this understanding of positive freedom as this emerges from the intersection of liberty, equality, and community. Rather than a necessary evil that must be limited in every way possible under the premise advocated by such thinkers as John Stuart Mill and Henry David Thoreau asserting "that government is best which governs least," positive concepts of democracy and community understand that government can really function as the empowering foundation of liberty, equality, and fraternity. *Government of the people, by the people, and for the people* is not only possible but necessary if we are to have authentic democracy. Government must arise from the people through the public space of dialogue and debate and a commitment to the common good.

Barker argues that "discussion" is the very essence of democracy. For when genuine political discussion takes place within society, individual points of view are modified, enhanced, and transformed toward a common good transcending the multiplicity of conflicting individual goods:

> So far as the society exists by dynamic process, it exists for and by the mutual interchange of conceptions and convictions about the good to be attained in human life and the methods of its attainment. It thus exists for and by a system of social discussion, under which each is free to give and receive, and all can freely join in determining the content or substance of social thought—the good to be sought, and the way of life in which it issues. Now such discussion is also, as we have seen, the essence of democracy. (1967: 19)

Similarly, Errol E. Harris observes that there is a "general will" or deeper "common good of the social whole" that arises when the community is understood as the matrix and womb of democracy rather than its nemesis. Human rights, as well as the possibility of a genuine common good, arise from within communities that form the matrix and sustainer of our individualities:

> Many thinkers who oppose civil society to the state continue the tradition of individualistic theories, which see law and regulation as a limitation on freedom rather than the means of its realization through the protection of legitimate rights.... It is from this mutual recognition alone that rights and duties derive, and it is only from this that one can decide what rights ought to be recognized, which (in any particular society) may not be. Apart from mutual recognition of persons there can be no prior "natural" rights attaching to individuals as such, as if they could exist independently of society and mutual recognition.... The democratic state, according to the thinkers we have been reviewing, is one in which the whole community participates, through its civil and administrative institutions, in a process of intercommunication, commerce and discussion, which generates a general will that, because it represents the common good of the

social whole, is supreme and thus exercises sovereign authority invested in the institutions of government. (2008: 32)

The concept of 'negative freedom' central to liberal democracy stems from the outmoded early-modern paradigm. Under the Lockean conception of liberal democracy, a conception largely institutionalized today in the United States and promoted worldwide through U.S. imperial policies, the individual right to unlimited accumulation of private property remains axiomatic, functioning within a conceptual framework in which government merely protects the competitive pursuits of private interests. Here the notion of the common good has little meaning beyond the idea of a common, conflictive pursuit of personalized goods. Immense concentrations of wealth give extraordinary political power to a tiny elite, undermining democracy at every turn. The second foundation of positive freedom, therefore, requires reasonable economic equality.

Karl Marx was fundamentally correct that political democracy cannot exist in any real way without economic democracy. A truly common good can only emerge when political democracy is supplemented by what Marx termed "substantive" or complete democracy. Economic democracy today must be understood, not as enforced collectivization (as the propaganda system would have us believe), but as the economic empowerment of individuals from the grass roots up in ways that supply the basic necessities of all persons while reducing vast accumulations of private wealth (equals power) that inevitably undermine and corrupt political democracy. Economic democracy does not abolish free markets, as we shall see. It simply prevents markets from being monopolized by the rich and powerful.

This conception of positive freedom emerging from genuine community does not involve a return to the religious or metaphysical foundationalisms of the past. It is rooted in the empirical realities of our situation much more firmly than the fragmented and negative conceptions of philosophical liberalism that derive from incorrect early-modern scientific assumptions. Traditional liberalism, pervasive still in the U.S. and many other countries, emphasizes autonomous individuals who have innate rights that put them over and against their neighbors and their governments. This liberalism is also a large part of the philosophy behind capitalism, seeing all individuals and corporations in egoistic competition with one another for scarce resources, profits, and ascendency, and seeing unlimited accumulation of private property as the center of their system of innate rights.

Many thinkers today are developing an alternative conception of positive freedom with its profound social and economic implications. Positive freedom links my freedom with that of everyone in the community. It understands that mutual cooperation and recognition of interdependency empower people and their life projects far beyond anything that liberalism can conceive. These thinkers include economists such as David C. Korten and Herman E. Daly, progressive writers such as William Greider and Naomi Klein, and philosophers such as Jürgen Habermas, Benjamin R. Barber, Enrique Dussel, and Errol E. Harris. Today, the "community" is everyone on the planet bound together holistically through democratic world law ensuring universal equity, prosperity, and sustainability.

At this juncture of planetary crises threatening the very future of human beings on this planet, the conversion to positive freedom takes on much more than theoretical significance. Early modern philosophies of atomism and consequent fragmentation have been overthrown across the board. We will only survive and prosper on this planet if the institutional and theoretical systems by which we conceptualize and organize our lives flourish in fundamental harmony with the scientific realities of our situation. We have moved beyond earlier metaphysical foundations for our human community, and we have moved beyond the atomism of early-modern thought. The new foundation involves the pervasive holism of our human situation as revealed by contemporary science.

Nevertheless, with the notable exceptions of Harris and Barker, most of the above mentioned thinkers do not see their way clear to embracing *planetary democracy*. They correctly understand that economics and government must be community oriented, rooted in localities, and organized in such a way as to allow moral and practical values of community members (winnowed through rational dialogue and debate) to articulate a common good that becomes the framework for action. Commodity and exchange relationships, the pure profit-motive, and government as a militarized security-system enforcing merely contractual relationships must be transformed into a framework in which people and nature come first, in which economics and government serve the common good and not the private interests of the few. Korten, for example, expresses this as follows:

> To survive on a living spaceship we must create a system of economic relationships that mimics the balance and cooperative efficiency of a healthy biological community. It must distribute rights equitably and link power to the consequences of its use. In short, we must commit ourselves to establish an economic, as well as political, democracy. A substantial body of proven experience suggests that such systems can be built around smaller enterprises functioning as self-directing members of larger networks in ways that are efficient, flexible, and innovative, and thereby secure the freedom and livelihood of the individual while nurturing mindful responsibility. (1999: 181)

This might serve as a brief description of meaningful positive freedom (which must include economic as well as political democracy) and what must happen on a planetary scale if we are to create a decent future for generations to come. However, like many of the other thinkers mentioned, Korten retains the liberal suspicion of government as a threat to liberty and does not deeply see the truth, emphasized by philosophers such as Hegel, Barker, and Harris, that the social matrix as a whole includes family, social organizations, and government as aspects of the same fundamental human community.

A radical distinction between government and civil society becomes less necessary within this framework (however true it remains that *vigilance* and political awareness on the part of citizens is the price of freedom). Korten realizes that the positive conception of democracy that he affirms must be planetary in scope. However, even in the face of the above enumerated planetary crises, he can only manage the weak and very unlikely suggestion that the current undemocratic domination of global trade and finances by the World Trade Organization and IMF be replaced with "an international agreement regulating international corporations and finance" (ibid. p. 191).

Korten's suggestion that sovereign militarized nation-states in historic competition with one another negotiate treaties for "an international agreement regulating international corporations and finance" becomes simply bizarre in the face of the global realities that threaten human existence on Earth. A similar criticism should be made of Barber's idea that citizens be consulted by "referenda" that replace representative national government as we know it with these new, cumbersome procedures. He calls this "strong democracy" and provides a compelling picture of what participatory governance would be like. However, rather than transforming government as we know it into authentic democracy and enlarging it to its proper global dimensions, he appears to reduce government to a truncated form that would be utterly incapable of dealing with the global crises that threaten humanity.

A similar lacuna appears in *The New Science of Sustainability: Building a Foundation for Great Change* by Goerner, Dyck, Lagerroos (2008), mentioned above. This excellent book delves deeply into the contrast between the older "world hypothesis" deriving from early modern science, liberal theories of democracy, and monopoly capitalism and the new science of sustainable economics, ecology, civilization, and community-oriented democracy. These authors see democracy as people empowered through cooperative, supportive social structures, vibrant communities, and horizontal economic relationships providing sustainable incomes for the majority. They critically examine neoliberal economics and its subversion of democracy through promotion of elite wealth and power at the expense of nature and the majority of persons.

Their book takes its stand on the need for "great change" through transforming all the failed assumptions of the traditional hierarchical concepts of modernity regarding democracy, economics, evolution, the structure of the universe, etc. The authors recognize the great transformation that contemporary science has effected: providing the insight that interdependency, unities, dynamic communities, and wholes sustain nature and the universe rather than the mechanistic, atomistic, and competitive relationships assumed by the early-modern hypothesis. However, amazingly, for all their claims that they are transforming the false set of modern assumptions, these authors *never* mention the sovereign nation-state.

They never recognize this institution as a foundational modern assumption creating fragmentation and war and wreaking havoc in our world. Their book includes an extensive analysis of capitalist economics and proposes economic democracy, but it omits the insight, expressed by social scientist Christopher Chase-Dunn, that "the state and the interstate system are not separate from capitalism, but rather are the main institutional supports of capitalist production relations" (1998: 61). Their book fails, therefore, to adequately lay the foundations for a sustainable civilization because they make their numerous arguments for great change without recognizing a central impediment to that change in sovereign nationhood.

Their analysis of democracy similarly fails on this account. They see the need for economic democracy, a "committed, collaborative community," "restoring civil society and civilizing mores" and "linking governance systems at multiple levels" (229-243), but they never follow through on the grounds of their own planetary perspective to the conclusion that "great change" requires planetary democratic government replacing the destructive system of autonomous sovereign nation-states that defeats planetary democracy, peace, and civilized human relationships on every side.

"Governance" cannot be linked at multiple levels, as they claim, if sovereign nations remain incommensurable with one another, each a law unto itself with no common law for them all. In line with these same uncriticized premises, the analysis of these authors focuses on the United States, as if creating holistic democracy in the U.S. will somehow solve our planetary crises and create a sustainable future for the Earth. The foundations they lay for sustainability appear seriously incomplete since they quietly accept a major impediment to sustainability inherited from false early-modern premises.

Goerner, Dyck, and Lagerroos promote the proper insight regarding our unity on the Earth, as when they quote R. Buckminster Fuller: "We are not going to be able to operate our Spaceship Earth successfully nor for much longer unless we see it as a whole spaceship and our fate as common. It has to be everybody or nobody" (230). However, they do not follow through on this insight to the conclusion that we must unite together in one democratic world order, that it *cannot* be "everybody" unless everybody is protected equally by a set of enforceable human rights and democratically legislated, enforceable laws. There is absolutely no way to include everybody unless *everybody* is united under a single constitution providing the framework for genuine planetary democracy. Their vision remains fragmented by incommensurables.

But positive freedom opens up new possibilities that are closed to negative freedom that is always in struggle with its perceived threats of government and community. By uniting as a global community under a democratic constitution, our positive freedom to deal with our planetary future would be immeasurably enhanced. Charles M. Sherover describes this process enhanced by positive freedom:

These positive freedoms or prescriptive rights are, be it

> noted, future oriented; they open up an area of futurity for
> exploration and development by individual members who
> may wish to do so, in the belief that such individual activ-
> ity contributes to the common good of all.... The specific
> positive freedoms or prescriptive rights that characterize
> the members of a particular society are their specified au-
> thorization to plan on a future opening before them; these
> positive freedoms or specified rights open up the vistas, as
> protected opportunities, their members, as a socially con-
> stituted body, authorize or encourage each other to pursue.
> (1989:79-80)

Our global situation today requires the vision to create a con-
stitutionally mandated planetary community of positive freedom
in which human beings are empowered to deal with the global
crises that threaten our existence and future generations on this
planet. Under our fragmented contemporary political and economic institutions, our freedom as human beings to create a sustainable future of peace, justice, and prosperity is severely restricted. Despite the best efforts of thousands of concerned world
citizens, little progress is made due to the security, economic self-
interest, and warlike measures of militarized nation-states with
their immense restrictions and practices that inhibit the actualization of a planetary common good.

The "common good of all" needs to be the product of a global
positive freedom that is only possible under a democratic *Earth
Constitution*. As Sherover suggests, the creation of government
alone will not suffice. However, government as the mainspring of
positive freedom that empowers and enhances the creative possibilities of the world's citizens remains the necessary prerequisite
for a truly transformed world order. Unless we *plan the future* together with the rest of the human community, there is little hope
for actualizing a future beyond our present suicidal trajectories.

Although Ernest Barker primarily restricts his discussion of
democracy to the nation-state, he does recognize the inherent universality of the concept of a democratic constitution as an "organization" that encompasses all men: "Ultimately, all other organiza-

tions of men must come to the bar of the organization of all men, if that can ever come to pass. We can imagine a high measure of general liberty under a system of national societies and national States. We can imagine a perfect liberty only in a world society and a World State" (1967: 28).

Hence, with the exceptions we have noted of Harris and Barker, most of today's thinkers concerning world problems mentioned remain trapped in the mythology of "sovereign nation-states" that history shows clearly as a major source of the global crises of war, poverty, exploitation, and chaos that we face today. As world citizen Harold Bidmead puts it, "The enemies of a peaceful international order are the worshipers of the false god of national sovereignty, the idol with feet of clay" (1992:123). My *Ascent to Freedom* (2008) traces the history of this mythology in some detail. Democracy must indeed be "strong democracy" empowered from the grass roots level, but democracy must be simultaneously *global democracy*, something that is only possible if sovereign *nation-states* federate as interdependent regions within a system of binding world law legislated by a genuine world parliament. This act, and this act alone, can open up a future for humanity.

The system of sovereign nations recognizing no effective law above themselves must be transformed into an Earth Federation legislating laws binding on all individuals and capable of ascertaining, implementing, and encouraging action for the common good of human beings and the Earth that sustains us. Twenty-first-century realities show that the system of sovereign nation-states has lost its raison d'être and no nation can any longer claim full democratic legitimacy. In *21st Century Democratic Renaissance* Harris writes:

> At the present time, the two fundamental principles justifying national sovereignty—the rule of law and the pursuit of the common good of its subjects, have become undermined, so that national sovereignty, as such, has become obsolete. Democracy, the sovereignty of the people, can no longer be claimed by separate national groups, because, as such, they can no longer maintain these two fundamental

conditions.... It is thus clear that under the currently existing world conditions the traditional conception of democracy, whether in the form envisaged by the individualistic
thinkers of the 17th to 19th centuries or that entertained
by later idealistic philosophers, cannot be realized. What
they conceived as the necessary condition of the exercise
of sovereign power in the state—its service of the common good—has become unsustainable within the national
state.... In the world today the only form of democracy that
could aspire to the ideals of the traditional philosophical
conception would have to be global, one that could legis-
late to implement global measures to deal with global problems (as sovereign nation-states cannot) and could maintain
the Rule of Law worldwide (which the exercise of sovereign
rights by independent nations prevents).... Were such a
democratically constituted World Government to be established it would not automatically solve the problems so urgently crying out for resolution, but it is the indispensable
precondition of any effectual remedy to global crises, if only
because sovereign nations are bound by their very nature
and definition to generate the conditions that exacerbate
current crises. (2008: 134-135 & 138)

It would not be prohibitively difficult to activate a global
grass-roots-oriented system of participatory democracy generating an authentic general will toward the common good. There are
already dozens of groups meeting and organizing in such a fashion to dialogue concerning global problems: from United Nations
conferences on the environment to the World Social Forum to Ox-
fam to Greenpeace to Amnesty International to the Provisional
World Parliament. And the computer technology exists that could
activate real global democracy. *The only legitimate sovereignty of the
people is that of all the people of Earth, for all persons on Earth have
the same universal rights, dignity, and equality and the same universal
responsibility for the future of the Earth and their children.* There is
no other route to generate a democratically articulated common
good of the Earth than through the universal participation of the
people of Earth.

The world community is now the only legitimate locus of sovereignty and the appropriate source of strong democracy. As long ago as 1946, Emery Reves stated that "as the world is organized today, sovereignty does not reside in the community, but is exercised in an absolute form by groups of individuals we call nations. This is in total contradiction to the original democratic conception of sovereignty" (132-133). The original democratic idea is that sovereignty resides in the community of the whole, Reves says, and the 20th century has realized that all humankind constitutes that community. The community of the whole then delegates its authority to the world, national, and local levels:

> Only if the people, in whom rests all sovereign power, delegate parts of their sovereignty to institutions created for and capable of dealing with specific problems, can we say that we have a democratic form of government. Only through such separation of sovereignties, through the organization of independent institutions, deriving their authority from the sovereignty of the community, can we have a social order in which men live in peace with each other, endowed with equal rights and equal obligations before law. Only in a world order based on such separation of sovereignties can individual freedom be real. (1946: 139-140)

This understanding properly turns on its head the notion that nations would delegate part of their "sovereignty" to world federation. Sovereignty, as the legitimate authority for planetary democracy, only derives from the community of all the people living upon the Earth. Nations can no longer be sovereign because, at least since the 19th century, the Earth has been understood as the planetary home of all humankind. Sovereign nation-states today exacerbate our planetary crises and destroy human freedom, which can only arise through a planetary community structured as a democratic federation dealing with problems at the local, national, and planetary levels. Neither can there be a common good for the planet arising from the general will of humanity without an Earth Federation under a genuine world constitution.

Jerry Tetalman and Byron Belitsos (2005) affirm this same truth today. They call national sovereignty "the profound political problem of our time" (22). They argue that "if sovereignty has its source in the people, and if the world has progressively moved in the direction of increasing democracy in recognition of this fact, then this concept must have an even greater destiny than we see today" (10). Sovereignty lies with the people of Earth, with the human community as a whole. They conclude that *"competitive nationalism is the greatest barrier to redefining our community as all humanity"* (15).

The sovereign people of Earth properly delegate some of their authority to governments: to local government to deal with local problems, to regional or national government to deal with regional or national problems, and to world government to deal with global problems. In the absence of the latter not only is positive freedom impossible, but community and democracy disintegrate as well. By abjuring the holism of our situation, we exacerbate our present fragmentation at every level.

To think there might be a democratic participatory formulation of a general will to a planetary common good simultaneously among some 193 sovereign nation-states is naïve utopianism of the worst kind. The strong democracy necessary to our present planetary situation is inhibited by the system of sovereign nation-states with their militarism, secrecy, nationalism, and competitive economics. In point of fact, the absolute sovereignty of nation-states has been steadily mitigated as the world has become more interdependent and as international law has evolved that recognizes universal principles applying to all states and persons, as within the new International Criminal Court (ICC).

However, this slow evolutionary process is entirely inadequate to deal with the nexus of interdependent global crises. The Earth will be practically uninhabitable by the time the U.N. evolutionary process has evolved into something approximating strong democracy. On the other hand, strong democracy can be activated with relative ease if these *impediments to democracy* are removed

and the *unity in diversity* of the peoples of Earth is institutionalized under a coherent *Constitution for the Federation of Earth.* Let
us step back and take a deeper look at the historical movement
that has led through several paradigm shifts to the realization of
holism as the final civilizational paradigm for planet Earth.

Chapter 2

The Historical Roots for Changing Paradigms

2.1 The Axis Period in Human History

THE Axial Period was a worldwide phenomena that gave birth to our present level of self-reflective consciousness. World civilization (and Western civilization) has developed, since the Axis Period of history (roughly the eighth to second centuries BCE), through three macro-stages or paradigms and many sub-stages. I will call the three central paradigms of this development the *Age of Static Holism* (characterizing many of the ancient civilizations through the medieval period) the *Age of Fragmentation* (characterizing the early-modern paradigm emerging from Europe and spreading worldwide after the Renaissance), and the *Age of Evolutionary Holism* (the global transition taking place today). The latter age has only recently emerged out of the discoveries of 20th century science and has not yet substantially influenced the world's dominant institutions (nation-states and global capitalism). However, from the understandings of our human situation in relation to the cosmos put forward during this age (holism), it is but a very short step to comprehending the necessity of demo-

45

cratic earth federation.

None of this is, so to speak, 'ancient history'. For Plato and the Buddha lived just yesterday on the scale of human existence on planet Earth (one to two million years), and they worked just two seconds ago on the scale of life on our planet in what biologist Loren Eiseley (1959) called the "immense journey" of development for the past 3.6 billion years. Still again, Plato and Buddha live as our virtual contemporaries on the scale of the age of the universe that exploded in its "primal flaring forth" some 13 billion years ago (Swimme and Berry 1992).

It was during the Axis Period of human history that philosophy, ethics, and advanced religion first emerged as an activity of reason attempting to self-consciously understand our human situation. Human self-consciousness had emerged to the point where reflection became inevitable and natural. This reflection was applied to ethics, law, social order, the natural order of things, transcendent reality, and nearly every other fundamental mode of human existence. The general tenor of the philosophies and religious world views coming out of the Axial period was holism.

In many parts of the world this ancient form of holism persisted nearly down to the present, with the notable exception of western society from the 16th century on, which introduced a philosophy of mechanism, atomism, and fragmentation, a philosophy that has since spread worldwide in the form of global capitalism and the nation-state system. My thesis includes the idea that human self-consciousness has continued to emerge since these early beginnings in the Axis Period. In the recent century we encountered a decisive breakthrough, a paradigm shift, not yet fully comprehended by our civilization. We began to understand the nature of the holism (inseparable from the parts) that is the dynamic and foundational order of our physical universe, biology, and human life.

The new holism is evolutionary. The universe and human life are no longer characterized by a static holism but by an evolutionary holism in which things change and develop through dynamic interaction with one another and with the wholes of which they are a part. The wholes or fields, that exist within ever-larger wholes or fields, provide a *telos* for the parts and their interaction. Human consciousness has evolved out of its primitive beginnings in the same way. Our thinking itself moves toward an ever-greater, dynamic evolutionary holism.

It is palpably obvious that human self-consciousness and reason have emerged out of the process of evolution on this planet. For most of our one to two million years there are few signs of self-awareness expressed in culture or artifacts. Not until about 15,000 to 40,000 years ago do we find major signs of the human symbolic capacity and imagination indicating an emergence of human consciousness from the homogeneous oneness of nature. This "magical" consciousness (as some have called it) is transformed into the "mythological consciousness" after the discovery of agriculture about 10,000 to 12,000 years ago. It was only in the "Axis Period" of the first millennium BCE that our species as a whole began to distinguish between our subjective responses to things and the objective way that things exist independently of those responses (Cf. Swimme and Berry 1992; Armstrong 2007; Hick 2004; Jaspers 1953).

The idea of an Axis Period has been discussed by many thinkers as a turning point in human evolution in which fundamental qualities emerged that define our current mode of consciousness. The transformation took place during the first millennium BCE all over the planet, but has sometimes been narrowed to the period from about 800 to 200 BCE. Karl Jaspers (1953) termed this age that of *Achsenzeit*. During this period great ethical and religious teachers arose who either gave birth to the great world religions of today or transformed existing religions into the more sophisticated forms that exist today.

During this period Lao Tzu, Chaung Tzu, and Confucius flourished in China. The *Upanishads* and *Bhagavad Gita* were written in India. The Lord Buddha founded Buddhism. Zoroaster lived and taught in Persia. A number of great Hebrew prophets lived who transformed Judaism. In Greece and Mediterranean Greek colonies, Pre-Socratic philosophers gave birth to philosophy and were followed by Socrates, Plato, Aristotle, and the early Stoics. Everywhere these teachings manifest a new sense of individuality and individual responsibility in relation to religion, to ethical action, and to human knowledge.

Some people were now capable of acting according to abstract ethical principles, not mere social custom and conformity directed toward preserving the mythologically grounded community. Oral traditions that had remained unchanged for centuries are suddenly open to examination from religious, ethical, and natural philosophers who are now looking for principles behind custom and nature, no longer just simply accepting the tradition. Those often considered the three earliest Pre-Socratic philosophers were Thales, Anaximander, and Anaximenes of Miletus in Asia Minor. Thales was teacher of Anaximander who was teacher of Anaximenes. It is not insignificant that each pupil disagreed with his teacher. The quest for knowledge of the cosmos grounded in objective principles had begun.

Individuals were first developing the ability of thinking for themselves and of acting autonomously. For the same reason, they are responsible for their ethical behavior (according to the great ethical and religious teachers of this period), which must now be based on abstract principles (such as the golden rule, see Hick 2004: Chap. 17), not mere custom and social conformity. And religion is becoming concerned with individual responsibility and salvation, not simply maintaining the unchanging human and cosmic order of things. Human consciousness has opened up to its own possibilities in relation to the ground of Being, ultimate reality, or God, and human beings are generally held responsible to transform their lives and the world according to a radical potential arising from the depths of reality.

In *An Interpretation of Religion: Human Responses to the Transcendent* (2004), John Hick studies this age in relation to the birth of the great religions. He writes concerning the mythic basis of the earlier, mythological age:

> This serves the social functions of preserving the unity of
> the tribe or people within a common world-view and at the
> same time of validating the community's claims upon the
> loyalty of its members. The underlying concern is conserva-
> tive, a defense against chaos, meaninglessness, and break-
> down of social cohesion. Religious activity is concerned to
> keep fragile human life on an even keel; but it is not con-
> cerned, as is post-axial religion, with its radical transforma-
> tion. (23)

We are witnessing the emergence of individuality, Hick says,
out of an earlier mode of human awareness not capable of clearly
separating its individuality from the group and its general environment. Here "the symbolization of self and world are only very
partially separate" (32):

> They were now able to hear and respond to a message relat-
> ing to their own options and potentialities. Religious value
> no longer resided in total identification with the group but
> began to take the form of a personal openness to transcen-
> dence. And since the new religious messages of the axial
> age were addressed to individuals as such, rather than as
> cells in a social organism, these messages were in principle
> universal in scope. (30)

These several features of the emergent consciousness of human beings during the axial age form an interrelated part of the
same transformation: the emerging ability to separate one's personal subjectivity from objective principles, an emerging individuality that distinguishes itself from the social group, an emerging sense of personal responsibility and personal potentialities,
an emerging sense of being able to live according to universal, abstract ethical principles, and an emerging ability to relate one's life
to the ground of Being or God, that is, to the whole of existence.
Human beings were emerging from their two million year evolutionary process into what I have called *rational freedom oriented
toward wholeness* (Martin 2008).

Some Pre-Socratic philosophers, living in Mediterranean Greek colonies, insisted that there was a special capacity within human beings that was to be distinguished from the other ways through which we responded to the world. This capacity was reason. Heraclitus identified it as the *logos*. The reason within had the ability to reflect or mirror the reason or *logos* hidden at the heart of the world process, the objective order of things. The human microcosm could rationally and objectively know the cosmic macrocosm. The older mythological consciousness that primarily consisted in a *response* to the world, in a relation to the world as a Thou, was coming to an end in an orientation that could *know* the world independently of the subjective feelings and responses of individuals or cultures.

The point that is most fundamental here is that the emergence of these qualities laid the foundation for democratic world law and, ultimately, sustainable civilization. Democracy is, most fundamentally, not simply one political system among others. It is an ethos and a set of values that correlate with the fundamental characteristics of human beings as these emerged during the Axis Period. These characteristics are still developing toward maturity within humanity. They are both qualities we inherent from transformations of the Axis Period and potentialities within us for ever greater realization.

The set of values that we call "democracy" arises from what we are as human beings as this emerged during the Axis Period: our sense of individuality, the sense of our human potential, the capacity for ethical responsibility according to abstract principles (fundamentally the ability to formulate and self-consciously follow laws), the ability to reason with all that this implies for knowledge and freedom, the ability to separate the merely subjective and personal from the objective, and the ability to progressively formulate and know the principles of nature (philosophy and science). These values simply state that society should be organized in such a way that these characteristics of our humanity are recognized, encouraged, built upon, and developed.

Democratic world law is inherent in our human situation since the emergence of these basic capacities that define our humanity. They still characterize our humanity today and define our potential for ever greater maturity. Since these capacities are universal they need to be institutionalized in a planetary civilization. A truly democratic world order will also mean a sustainable civilization. Economic and political relations under democracy are not fragmented into militarized antagonistic units, nor are they fragmented into exploiters and exploited. As we will see, democracy implies all three dimensions of liberty, equality, and community on a planetary scale. A sustainable, holistic civilization will be the result of the ascent to authentic planetary maturity and democracy.

Democracy must be founded on the principle of *unity in diversity*. It must be holistic in that the diversity of individuals must be encompassed within a holistic framework recognizing the equality, species sameness, rights, and dignity of all members of the society. It is here that early modern physics and philosophy took a wrong turn that has such serious consequences for our endangered world-order today. For it could not see its way clear to any serious form of holism. The science that emerged from the 17th and 18th centuries appeared to reveal a physical and social world-order of fundamental diversity (fragmentation) without any significant redeeming unity.

2.2 The Paradigm of Newtonian Physics and Philosophy

This separation of subject from object during the Axis Period eventually made possible the discovery of the scientific method during the 17th century. The observer created an ideal situation called an "experiment" to test a hypothesis under controlled conditions. The observer was a detached spectator who observed the results in quantifiable and systematic terms. The result was an explosion of knowledge about the physical world. Neither subjective feelings nor the observer's values were to interfere in the observations. Fact and value were considered two different dimensions of existence.

In 1687, Sir Isaac Newton synthesized the laws of planetary motion discovered by Johannes Kepler with the laws of earthly motion discovered by Galileo Galilei. It appeared as if science had uncovered the most fundamental workings of the universe. Newton posited an absolute space and time within which the mechanics of bodies in motion operated. The universe was conceived on the model of a vast machine whose parts (bodies) operated in external relations with one another and which were reducible to simpler parts down to the smallest atoms.

In the 17th century, René Descartes posited the mind as a "mental substance" entirely different from matter or "extended substance." Mind and matter confronted one another, one physical and the other non-physical. There seemed to be no easy bridge between them. The question of how non-physical mind could control physical matter (which it obviously did in the human body) led Descartes into great conceptual difficulties. However, subject and object were now formally distinct, and the project of knowing the world "objectively" could proceed unhindered. With their connection so severed, Descartes (in his Sixth Meditation) ultimately had to depend on God as a *deus ex machina* to guarantee the veracity of the mind's relationship with the external world (1975:185-199).

All the early-modern thinkers were strongly influenced by the assumptions of the newly emerging science of physics, from Spinoza, Leibniz, and Kant on the European continent to the empiricists, Locke, Berkeley, and Hume in England. In his *Essay on Human Understanding*, John Locke attempted to construct human knowledge on the basis of "simple ideas" that entered into the "blank tablet" of the mind through the five senses. As an empiricist, he would only accept ideas that originated in sense experiences of the primary and secondary qualities of bodies in the world. However, he soon realized that one never experiences the "substance" within which these qualities appear to inhere. He concluded that this substance underlying the qualities of the world was "something...I know not what" (1978: 101).

Bishop George Berkeley, also an empiricist, followed the Lockean path with respect to substances to its logical conclusion. If substances were not observable, they could not be part of human knowledge and must be dropped as a fiction. "To be means to be perceived" (*esse est percipi*), he concluded. All things existed as collections of ideas and these collections of ideas were available to perceivers, including the ultimate perceiver who guaranteed the solidity and stability of the world: God. Berkeley perceived that subject and object must be internally related in some way ("to be" must be internally related to a perceiver). But his Newtonian atomism and reductionism prevented his moving forward with this insight and required that, like Descartes, God be brought in as a *deus ex machina* to guarantee the existence of the world when no finite creatures were perceiving it (1957).

David Hume drew even more radical conclusions based on his powerful, logical empiricism. The "simple ideas" of Locke become "simple impressions" for Hume. And it seemed that there was no impression for the vaunted "universal causality" of Newtonian physics. In the relation of two events in which one is said to cause the other, we only perceive "constant conjunction," never necessity. Hence, necessity in nature must be dropped as a fiction. But neither do we perceive any unitary, subsistent "self" that Descartes had assumed as the objective observer of the world. If we don't perceive it, it too must be dropped (1962). The result was a skepticism so pervasive that Immanuel Kant realized that something had to be wrong. Kant said that Hume's philosophy had awakened him from his "dogmatic slumber" to a rethinking of the entire grounds of human knowledge.

Empiricism, as part of this early-modern set of assumptions, did not appear to be able to account for human knowledge. However, Kant's rethinking of the Cartesian starting point was hindered by the fact that he, too, assumed the fundamental principles of the Newtonian paradigm. Nevertheless, he understood that subject and object must be *internally related* in any coherent account of knowledge and developed his *Critique of Pure Reason* (1781) on this basis. And he understood that the "transcendental unity" of the subject of knowledge (in sec. B 139) is correlative to the demand of reason that we find systematic unity in the world. With this, Kant points forward to the 20th century scientific paradigm that has discovered the fundamental unity of the universe and the holism by which diversity is integrated into that unity.

The law of reason which discerns relations within ever-greater unities and requires us to seek this unity, Kant deduced, is a necessary law, since without it we should have no reason at all, and without reason no coherent employment of the understanding, and, in the absence of this, no sufficient criterion of empirical truth or any truth. In order, therefore, to secure an empirical criterion we have no option save to presuppose the systematic unity of nature as objectively valid and necessary(1965b, sec. B 679: 538). The revolutions in 20th century science have borne out Kant's deductions.

In these ideas of Kant we find the principles that were soon picked up by G. W. F. Hegel and became fundamental to a new paradigm that not only recognized the inseparability of subject and object in the constitution of knowledge and the holism of nature but added to this the understanding of *emergent evolution* that sees human life as capable of evolving into truly new paradigms. Thinkers like Kant and Hegel were discerning the flaws in the mechanistic and fragmented worldview of early-modern philosophy and science. However, it was too late—the basic premises of the early-modern paradigm had become fundamental not only to many philosophers, as we have seen, but were now embodied within worldwide institutions like capitalism and the system of sovereign nation-states.

In the case of each empiricist thinker (Locke, Berkeley, and Hume), "simple" ideas or impressions were treated like Newtonian atoms, and the attempt was made to construct human knowledge from these simples, just as the world was understood as a machine and said to be constructed from material atoms in external relationships with one another. The world was understood in terms of universal efficient causality only, and the teleology fundamental to both ancient and contemporary thought was excluded. The world appeared to operate like a vast machine through its own self-contained motions. God was relegated to the role of an original creator who no longer played a part in the functioning of the machine.

Similarly, modern capitalism, developing since the Italian Renaissance and theoretically formulated in the work of Adam Smith and others in the 18th century, treats human beings as if they were Newtonian mass-points. The multi-dimensional complexity of human beings is ignored in favor of a conception of persons as individual self-interested atoms seeking only to maximize their personal advantage. Just as the bodies (mass-points) of Newton were governed in their external relations with other bodies by causal laws within the framework of universal gravitation, so the capitalist paradigm imagines its self-interested human units within what it assumes to be natural laws of supply and demand.

And just as the bodies in motion of Newtonian physics meshed together in the wondrous orbiting of the planets and the marvelous intricacy of natural processes on Earth, so the human atoms of capitalism are inserted into mathematical formulas producing wonderful theoretical benefits for the majority of humanity. Adam Smith first formulated this seemingly miraculous outcome as an "invisible hand." The meshing of innumerable greedy atoms in the "free" marketplace produces the greatest benefit to the greatest number of people.

The fact that this economics had to be (and continues to be) forced on the world through brutal colonial and imperial wars, and that no nation or corporation has ever followed or believed in "free trade" (what they want is control of markets eliminating competition), and that poverty and misery in the world have increased everywhere capitalism has extended its greedy hands, are entirely lost on the legions of academic mandarins and philosophers of exploitation who continue, to this day, to spew forth versions of this classical ideology (cf. Smith 2005, Ch. 6). The fragmentation of capitalism has to be enforced through a fragmented system of sovereign nation-states also based on the early-modern mechanistic model that assumes the self-interested competitive orientation of each of the atoms in the system. Each nation operates out of self-interest, just like the hypothetical individuals and corporations of capitalism operate out of competitive self-interest.

Newtonian physics is mirrored in early-modern social theory, economic theory, and philosophy. It has continued into 20th century positivism, logical empiricism, and so-called "epistemological realism." It has also continued into 20th century capitalist economics and nation-statism. It is reflected in the "political realism" theories put forward by thinkers like Hans Morganthau (1993) or Leo Strauss who saw the nation-state system as an amoral struggle of giant power centers for ascendency and hegemony. This political realism still dominates the foreign policy of the United States (cf. Engdahl 2009, Hardt and Negri 2000). It is reflected in a different way in the "world systems theories" of thinkers like Immanuel Wallerstein (1983) and his many academic followers who appear to see no hope beyond the perpetual struggle of the global imperial centers with one another for control of markets, resources, and exploitable labor (cf. Shannon 1989).

2.3 The Paradigm Shift of the 20th Century

The distinction that emerged during the Axis Period between our subjective perceptions and the objective world became the basis for the ancient, medieval, and modern paradigms that can be roughly described (in spite of many individual variations) in the development of human history. 19th century sociologist Auguste Comte also saw history as progressive. He formulated three large stages in historical development that he labeled the theological, the metaphysical, and the positive. The theological included both ancient and medieval thought; the metaphysical included the 17th, 18th, and 19th century philosophies that appealed to ultimate, untestable principles such as democracy, human rights, or sovereign states as autonomous personalities. The positive age involved the restriction of knowledge to what is confirmable by science.

Under the age of science, Comte believed that real progress was possible from material progress, to improvements in human physical well-being, to intellectual progress in understanding scientific knowledge and its foundations, to moral progress that included increased common sympathy, benevolence, and sense of community (Blain 2004b: 16). Comte's relatively short view of development covering only the past 2500 years or so can be questioned on empirical grounds. His basic idea (that moral progress is possible and emerging from the process itself) can be placed within the context of the 1-2 million years of human existence on this planet. Only yesterday (on this time scale) did we begin to think in terms of moral principles, right and wrong, and society in relation to these issues. Today, we see that scientific and technical progress are not equivalent to a progress in human well-being. Nor are they equivalent to progress in human moral consciousness.

Progress to date is far from obvious given that human beings remain mired in the outmoded Newtonian paradigm in terms of economics, nation-state political organization, and social theory. But the paradigm shift of the 20th century does point forward to a new foundation for economics, government, and social theory. The new paradigm does not replace the older Newtonian paradigm with an incommensurable alternative, as Thomas Kuhn (1970) would have it, suggesting a relativism of paradigms. Rather, according to physicist Henry Stapp, the Newtonian view was *"fundamentally incorrect"* (in Kitchener 1988: 56). The new paradigm encompasses the older within a larger and more coherent framework, a framework revealed on every side by 20th century breakthroughs in science, and replaces those assumptions that were incorrect.

Philosophic foundations derive from an examination of the coherence of experience on every side. Strict empiricism is an epistemology that derives from the Newtonian paradigm and assumes that knowledge is built up from discrete sensations organized into patterns by self-conscious observers. However, such reductionistic and atomistic empiricism cannot begin to account for 20th century developments in human knowledge and understanding of the holistic universe in which we live. Observation, of course, plays a necessary role, and we are far removed from any rationalist reliance on "pure reason." But the criterion of truth is coherence of experience with thought, and this coherence reveals a universe of indubitable wholeness encompassing a multiplicity of parts in a descending series of "fields" or systems. It reveals a universe observably structured on the principle of *unity in diversity*.

Human beings have emerged out of a cosmic process perhaps thirteen billion years in the making. This cosmic process appears to us as an comprehensive whole integrating a multiplicity of elements into ever-greater levels of complexification resulting in ever-greater levels of *unity in diversity*. Human beings exemplify the highest level known to us of the complex integration of *unity in diversity*. We are self-conscious of our situation as one species inhabiting planet Earth, and we are conscious of our vast multiplicity of unique differences among individuals, ethnicities, cultures, and religions.

Among many advanced thinkers, our emergence out of the cosmic process has transformed the way we look at ourselves. Advanced psychologists, physicists, and philosophers have been articulating the new holistic paradigm throughout the 20th century. But this new paradigm, premised on emergent evolutionary development of *unity in diversity*, has not transformed general thinking, nor our outdated institutions. Physicist Henry Stapp compares the view of our selves derived from "the classical view" of early modern philosophy and physics with what he takes to be the substantially correct view put forward by physicist Werner Heisenberg.

> This classical view of man and nature is still promulgated in the name of science. Thus, science is seen as demanding a perception of man as nothing more than a local cog in a mechanical universe, unconnected to any creative aspect of nature. For, according to the classical picture, every creative aspect of nature exhausted itself during the first instant....
>
> In the Heisenberg ontology, the real world of classical physics is transformed into a world of potentialities, which condition, but do not control, the world of actual events. These events or acts create the actual form of the evolving universe by deciding between the possibilities created by the evolving potentialities. These creative acts stand outside space-time and presumably create all space-time relationships. Human mental acts belong to this world of creative acts, but do not exhaust it. (In Kitchener 1988: 56-57)

The classical view of nature exhausted creativity in the first instant because a mechanism is fixed by its structure and understood in terms of efficient causality rather than a creative *telos* guiding the emergence of what is genuinely new, the realization of potentialities that are not yet actualities. Human beings are now understood as an integral part of the whole because science has discovered that everything in the universe is related to everything else, and that the evolutionary process itself cannot be understood in terms of the mechanical "Darwinian" mechanism of random mutations in genetic material giving rise to anomalous characteristics that are then selected out through a struggle for survival among atomistically conceived creatures (Darwin himself never put it in terms of this model).

Stapp asserts that potentiality is prior to actuality, that the universe is an emergent process of realizing higher potentialities, not a mechanical process of evolution pushed from behind by efficient causalities. Further, for Stapp, the human mind participates in the creative process that is at the heart of the universe itself, a perspective that coheres with the view of many psychologists, outlined above, that a human being must be understood in terms of the primacy of his or her potentialities. Stapp asserts that the potentialities themselves *evolve*. New conditions in the universe give rise to new possibilities.

If this is the case, those who claim that democratic world law and a sustainable world system will take decades or centuries to evolve are still thinking in terms of the social relationships and causal mechanisms derived from the early-modern paradigm. The global crises that confront humanity put a pressure upon human consciousness and simultaneously create potentialities that did not exist a mere fifty years ago. Rapid transformations of consciousness, attitudes, institutions, and relationships are very possible. The possibility of democratic world law is not fixed in the causal relationships derived from outdated institutions. It dynamically emerges from within the process and may come to fruition very rapidly. We will see below that a civilization under democratic world law can be *founded*. We need not wait for it to slowly evolve.

In his book *The Tao of Physics* (1975), physicist Fritjof Capra describes the interconnected holism of our universe that has been the central discovery of 20th century physics:

> Thus modern physics shows us once again—and this time at the macroscopic level—that material objects are not distinct entities, but are inseparably linked to their environment; that their properties can only be understood in terms of their interaction with the rest of the world. According to Mach's principle, this interaction reaches out to the universe at large, to the distant stars and galaxies. The basic unity of the cosmos manifests itself, therefore, not only in the world of the very small but also in the world of the very large; a fact which is increasingly acknowledged in modern astrophysics and cosmology. In the words of the astronomer Fred Hoyle:"Present-day developments in cosmology are coming to suggest rather insistently that everyday conditions could not persist but for the distant parts

of the Universe, that all our ideas of space and geometry would become entirely invalid if the distant parts of the Universe were taken away. Our everyday experience even down to the smallest details seems to be so closely integrated to the grand-scale features of the Universe that it is well-nigh impossible to contemplate the two being separated." (209-210)

Once again, the holism of the universe is identified, and with it the dynamic interdependence between local conditions and macroscopic conditions. Human beings, too, are implicated in this holism, and it is no anomaly that the idea of an "anthropic principle" originated among physicists themselves. This is the idea that the emergence of human beings has been integral to the conditions and emergent possibilities of the universe from its very beginning. If the holism of human life on this planet clearly dawns upon us, democratic world law and sustainable civilization will soon follow.

Philosopher and scientific cosmologist Milton K. Munitz, in his book *Cosmic Understanding: Philosophy and Science of the Universe* (1986), remarks on the "Anthropic Principle" developed by these 20th century physicists. Physicists have discovered the astonishing connection between the exact physical parameters of the universe and the emergence of human life:

That cosmology might provide such a fresh perspective is related to recent discussions of the *anthropic principle*. One version of this principle calls attention to certain special circumstances of a cosmological character necessary for the very existence of life. When examined in this perspective, the anthropic principle undertakes to make clear why it is that without these special cosmologic conditions, life would not exist at all. The very same principle, however, has also been used in order to show that the fact of human existence illuminates special features of the actual universe as compared with other possible universes. It is not simply that certain properties of the universe throw light on human existence, but the reverse also holds. The universe and human life are coupled. If we are to understand either, we need to move in both directions: from the universe to man and from man to the universe, since they are mutually involved in a very special way. (236-237)

It is high time that philosophers of law, jurists, and legal theorists begin appropriating the transformed paradigm bequeathed to us by 20th century science. For democratic law is such a fundamental expression of our human reality that a theory of law based on false foundations, on mistaken premises, can only spell disaster for human existence on this planet. As Munitz puts it, "the universe and human life are coupled." We need planetary institutions premised on this connection.

The implications of the early-modern paradigm were discerned, perhaps most clearly, by the 19th century philosopher Friedrich Nietzsche. He perceived this tendency toward disaster in the Newtonian science of his day. In the *Genealogy of Morals* (III, 25), Nietzsche expresses his dismay at the devaluing of human beings by the apparent implications of science:

> *All* science (and by no means only astronomy, in the humiliating and degrading effect of which Kant made the noteworthy confession: "it destroys my importance"...), all science... has at present the object of dissuading man from his former respect for himself, as if this had been nothing but a piece of bizarre conceit.

The presuppositions of much contemporary philosophy of law and government involve the Newtonian and Darwinian views of nature and human beings, views that reduced human beings to insignificant creatures struggling for survival in a vast, impersonal universe. The Newtonian view further assumed an atomism that viewed the universe as a collection of mass-points representing bodies in motion within a framework of absolute space and time. However, every one of these presuppositions has been transcended by contemporary science.

Menas Kafatos and Robert Nadeau, in their book *The Conscious Universe: Part and Whole in Modern Physical Theory* (1990) after explicating the emergent 20th century paradigm of the inseparability of part and whole, make this point in the following way:

> And yet it is also demonstrably true that theoretical reason does over time refashion the terms of construction of human reality within particular linguistic and cultural contexts, and thereby alters the dynamics of practical reason. As many scholars have exhaustively demonstrated, the classical paradigm in physics has greatly influenced and conditioned our understanding and management of human systems of economic and political reality. Virtually all models of this reality treat human systems as if they consisted of atomized units which interact with one another in terms of laws for forces external to the units. These laws or forces are also assumed to act upon the isolated or isolatable units to form hierarchical organizations which are themselves isolated or isolatable from other such organizations. (181)

20th century physics has discovered that human beings cannot be treated with this atomized approach because everything in nature is integrated into wholes and cannot be understood apart from the *unity in diversity* of holistic fields. This understanding necessarily alters the dynamics of human practical reason (our values and ethical actions)."Economic and political reality" must be transformed to appropriate the new holism of *unity in diversity*. According to Capra, the new holistic model in physics is mirrored across the spectrum of sciences whose paradigm now examines the world from the perspective of "self-organizing systems."

> The broadest implications of the systems approach are found today in a new theory of living systems, which originated in cybernetics in the 1940s and emerged in its main outlines over the last twenty years.... As I mentioned before, living systems include individual organisms, social systems, and ecosystems, and thus the new theory can provide a common framework and language for a wide range

of disciplines—biology, psychology, medicine, economics, ecology, and many others. (In Kitchener 1988: 149)

For Capra, self-organizing systems do not have their organization imposed upon them by external forces (the environment) as in the Newtonian model. Rather, they tend toward establishing their own order in interaction with the environment (ibid. 149-150). This approach means that nature includes within itself a *nisus* (internal self-direction) for self-organization and wholeness. Ultimately, this approach also includes the human mind. Mind is no longer understood as a mysterious opposite of matter as in the Newtonian modern paradigm. And it certainly cannot be understood as an egoistic atom of rationalized self-interest imagined by the theorists of capitalism. Rather, mind is now integral to "matter" at every level:

> The organizing activity of living, self-organizing systems, finally, is cognition, or mental activity. This implies a radically new concept of mind, which was first proposed by Gregory Bateson (1979). Mental process is defined as the organizing activity of life. This means that all interactions of a living system with its environment are cognitive, or become inseparably connected. Mind, or more accurately, mental process is seen as being immanent in matter at all levels of life. (Ibid. 151)

In the new psychology that we reviewed briefly above, mind can be expressed on various organizational levels and is not understood as different kind of substance from its environing universe. In his book *Human Potentialities* (1975), psychologist Gardner Murphy concludes that the natural human being and the human mind may well express "cosmic potentialities, cosmic trends as yet unparalleled elsewhere in the knowable universe." The traditional dualism between mind and matter is no longer tenable in a universe now understood as composed of various forms of organizational energy. Neither is fact any longer easily separable from value, since the holism of our situation no longer separates

human beings from the environing world as an independent reality examined by objective observers who somehow stand apart
from that reality. Yet our human potentialities remain "cosmic"
and enormous.

Our immense potential for a fundamental transformation to
holism must necessarily include every aspect of life on our planet.
The biosphere is holistic—and to survive and flourish human life
must become holistic across the board. We must not only deal
with the *scientific* realities of how the biosphere operates, we must
deal with the *spiritual* level from which each of us lives, the *cultural* dimension of attitudes, customs, traditions, etc., and the social *systems* through which we do business, by which we settle
differences, and within which nations and groups relate to one
another. Integral scholar and thinker, Ken Wilber, calls this "integral ecology":

> The basic idea is simple: anything less than an inte-
> gral or comprehensive approach to environmental issues is
> doomed to failure.... *Exterior* environmental sustainability
> is clearly needed; but without growth and development in
> the *interior* domains to world-centric levels of values and
> consciousness, the environment remains gravely at risk.
> Those focusing only on exterior solutions are contributing
> to the problem. Self, culture, and nature must be liberated
> together or not at all. How to do so is the focus of integral
> ecology. (2007: 100)

The new holism requires that we *think* as parts of encompassing wholes, *act as cooperative* parts of wholes in internal relationship to the wholes that encompass and sustain us, and that we
establish institutions reflecting this holism, which will, in turn, influence culture and human spirituality toward holism. It is in
the area of global institutions that most of the theorists discussed
above have failed. They assert that we need to begin thinking
as integral parts of a human community and our natural environment. They assert that we need an economics of cooperation, equity, and universal minimum prosperity rather than one

of exploitation, domination, and extreme differences of wealth and poverty. But they naively never seriously question the system of nation-states dividing humankind into 193 separate territories with absolute boundaries and national institutions of self-promotion and inter-state competition. They have not fully ascended to a holistic perspective, for such a perspective necessarily requires institutionalizing global democracy.

2.4 Creation of the Constitution for the Federation of Earth

We have seen that holism remains at the heart of most arguments put forward by environmentalists and ecologists, who assert that we cannot successfully achieve a sustainable civilization unless we begin to see ourselves as part of the natural world and our actions as integral to the functioning of the planetary biosphere. And we have seen above that this understanding does not eliminate human spirituality, religion, ethical universality, or any of our other 'higher' human functions. It merely reveals those higher functions as emerging from the mysterious depths and natural processes of cosmic evolution and natural evolution on our planet. We have also seen that what we need more than anything else, at this point in history, are institutions uniting human beings into a global community in which our higher human functions of reason, spirituality, and ethical universalism can flower into an effective, democratic decision-making process for our planet.

For these reasons, the *Constitution for the Federation of Earth* may well be the most important document of the 20th and 21st centuries. It is comparable in significance to the U.N. Universal Declaration of Human Rights or the Charter for the International Criminal Court. However, unlike the latter documents, the *Earth Constitution* will be hailed as *establishing the paradigm shift* that made possible peace, justice, and environmental sustainability for the Earth. These latter documents appear among those representing the highest moral and legal thinking possible under the present world order dominated by the system of sovereign nation-states for well over four centuries. The *Earth Constitution* establishes the foundations of a transformed world order premised on the *holism* that has been uncovered by every 20th century science from micro-physics to macro-physics to ecology to systems theory to the basic social sciences.

The world-wide movement for a government for the Earth began as a serious movement among thinking people during the First World War. It expanded between the world wars and became a very powerful world movement in the years following the Second World War. A group of people who were active promoters of a world parliament and world law at the time understood that the most compelling need for our endangered Earth was to have a completed constitution available for the people of Earth to ratify before it became too late. Philip and Margaret Isely, a husband and wife team based in Denver, Colorado, were among the leaders of this movement to get a quality constitution produced for the Earth. (The following history is based on Martin 2010, Chap. 1.)

In the mid-1950s, Philip and Margaret Isely joined the Campaign for World Government at its Chicago offices, at that time under the direction of Mary Georgia Lloyd. Along with Thane Reed, Guy Marchand, Marie Philips Scot, and others, a "World Committee for a World Constitutional Convention" was formed which, by 1961, established its headquarters in Denver, Colorado. The public call for a World Constitutional Convention was issued by the committee that same year with committed delegates from 50 countries and endorsements from several heads of state.

By 1966 the decision was made to change the name of the World Committee for a World Constitutional Convention to the World Constitution and Parliament Association (WCPA). Margaret and Philip Isely had been using the profits from their successful Denver-based business to travel widely, recruiting prominent persons to sponsor the development of a world constitution and prepare the call for a World Constitutional Convention. Among the recipients of Philip Isely's immense correspondence were Dr. T. P. Amerasinghe of Sri Lanka and Dr. Reinhart Ruge of Mexico, both leading world federalists who had independently arrived at similar conclusions. These activists eventually became Co-Presidents of WCPA and worked together for many years in this capacity, with Philip Isely as Secretary-General and Margaret Isely as Treasurer.

Three preparatory congresses were held in the mid-1960s, systematically building support and ideas for a world constitutional convention. The Convention, which took place in 1968 in Interlaken, Switzerland, and nearby Wolfach, Germany, drew 200 delegates from 27 countries and five continents. The Convention (now calling itself the First Constituent Assembly) formulated the major elements to be included in this constitution and elected a drafting commission of twenty-five persons, chaired by Dr. Reinhart Ruge, to complete a draft and circulate it worldwide for comment and criticism, setting the date of 1977 for its next meeting and the completion of this process. In his autobiography, Ruge writes:

> Wolfach was the real beginning of the attempt to create a stable world, which would save future generations from war and misery. This was all basically due to the clear line of thought of Philip Isely, and his capacity to find and bring together so many likewise intentioned people from around the world. I am very proud that I could be present at this important and historic Constituent Assembly. (2003: 305)

The year 1968 was truly an auspicious year in the struggle for a transformed world order. Boswell and Chase-Dunn in *The Spiral of Capitalism and Socialism,* speak of "the world revolution of 1968" (2000: 111) in which spontaneous rebellions erupted around the world against the old system of nation-state hegemony protecting a global economic system of domination and exploitation. The corrupt nature of the Democratic Party in the U.S. was revealed in the brutal repression of protesters at the Democratic Convention while spontaneous uprisings of students challenged the global order in Paris, Warsaw, Prague, and Mexico City. World renowned philosopher Emmanuel Levinas (2006) writes "in the fulgurance of certain great moments of 1968, quickly extinguished by a language just as wordy and conformist as the one it was supposed to replace, youth consisted in contesting a world already denounced long ago" (2006: 69).

While Levinas is correct that the "wordy" propaganda of the dominant system of state-capitalism quickly buried the dissenting voices under a barrage of propaganda, a new, truly democratic world order was being founded at Interlaken and Wolfach that went beyond simply idealistic words of protest to the creation of a founding document. A procedure was established for creating an *Earth Constitution* that could truly transform the world order by transcending verbal ideals in a concrete document subject to ratification by the people of Earth.

In 1972, five key members of this drafting commission met for two continuous months in Denver, Colorado, and created the first draft of A *Constitution for the Federation of Earth*. The following year this was circulated worldwide for comments and criticisms. In 1975, all these comments were collected and circulated worldwide, and in 1976 a second draft of the *Constitution* was prepared by the commission. This new draft was then also circulated worldwide as preparatory for the Second Constituent Assembly that met in Innsbruck, Austria in June 1977. At Innsbruck, this collectively revised draft for the *Constitution* was debated and amended paragraph by paragraph by the delegates. It was then adopted with 138 signatories from 25 nations and six continents.

In the following two years, the *Constitution for the Federation of Earth* was translated into a number of languages, sent to all Heads of State, and circulated widely. In response to a common criticism that no national governments had participated at Innsbruck, a Third Constituent Assembly met at the Hotel Ranmuthu in Colombo, Sri Lanka in 1979, hosted by WCPA Co-President, Dr. Terence Amerasinghe. This body did not find it necessary to amend the *Constitution*. Rather, the Assembly issued a Declaration of the Rights of People to assemble, draft a constitution, and obtain ratification. A key issue of world federalism (and important for the future of humankind) was thus delineated at this point. Does the future of the world lie entirely in the hands of illegitimate sovereign national entities militarizing the world and creating ever more weapons of death and destruction? Or do citizens of the Earth have the right and duty to take charge in creating a decent world order for themselves and future generations?

During the 1980s, the World Constitution and Parliament Association focused on organizing sessions of the Provisional World Parliament under the authority of Article 19 of the *Earth Constitution*. The Parliament met in Brighton, England, in 1982, Delhi, India, in 1985, and Miami Beach, Florida, in 1987. However, criticisms of small details in the wording of the *Constitution* kept surfacing to the point where it was deemed necessary to call one final World Constituent Assembly for 1991. This was held in Troia, Portugal, at which time the delegates adopted 59 (mostly small) changes in wording within the *Constitution* and renewed the worldwide campaign for its ratification, which was then called the Global Ratification and Elections Network (GREN) and later known as the Earth Federation Movement (EFM) (Cf. Martin 2011).

During the 1980s, the initial sessions of the Provisional World Parliament were quite successful. The first session in 1982, at the renowned Royal Pavilion in Brighton, England, attracted delegates from 25 nations and six continents. The impressive inauguration of the Parliament was presided over by Sir Chaudry Mohammed Zafrullah Kahn of Pakistan, who was former President of the U.N. General Assembly and former foreign minister for his country. Officers of the Parliament included such notables as Lucile Green (later President of the World Citizens Assembly), Max Habicht (renowned international lawyer), and A.B. Patel, then Secretary-General of the Sri Aurobindo Movement and World Union, headquartered in Pondicherry, India.

The Second Session in 1985 inaugurated before a packed house in the famous Constitution Club of Delhi (where the Constitution of India had been signed). It was opened by the then

President of India, Zail Singh, and chaired by the then Speaker of the Lok Sabha (the lower house in India's Parliament), the Hon Bal Ram Jakhar. The Third Session met at the elegant Fontainbleu Hilton Hotel in Miami Beach, Florida for eleven days of intense work during June 1987. Along with passing a number of important world legislative acts, it included an exposition for developing countries to show their products and wares and began the elaboration of the Ministries of Provisional World Government as sanctioned by Article 19 of the *Earth Constitution*.

During this decade, hundreds of organizations worldwide were committing support to the *Constitution for the Federation of Earth*. The heads of some of the poor nations were expressing interest and meeting with WCPA leaders. The campaign for ratification of the *Constitution* was in full swing with the signatures of personal ratifiers flooding into the Denver offices of WCPA, and a large network of WCPA chapters and organizational affiliations were developed throughout the world.

However, the end of the Cold War simultaneously cooled many people's sense of the necessity for ratifying an *Earth Constitution*. And, at that time, the great urgency to prevent climate collapse had not yet taken hold of the popular imagination. Today, more than two decades later, people again are taking great interest in ratifying the *Earth Constitution*, since they are beginning to understand that there is absolutely no way a world dominated by corporate capitalism and divided into 193 militarized nation-states can ever unite sufficiently to save our planetary biosphere from collapsing.

Chapter 3

A Global Social Contract for Sustainability

3.1 The Conceptual and Structural Holism of the Earth Constitution

THE Preamble to the *Earth Constitution* provides the conceptual framework for the whole of the document. It gives us the language of a "new world which promises to usher in an era of peace, prosperity, justice and harmony." Given the bleak and bloody history of humankind to date, how can the framers of the *Constitution* be so confident? The answer is given in the second paragraph of the Preamble: "Aware of the interdependence of people, nations, and all life." It declares the need for a sustainable planetary civilization. This is a declaration of holism that could not be clearer: there is no such thing as autonomous independence from the rest of humanity, from the other nations of the world, or from the natural world.

The next four paragraphs in the Preamble address the consequences of the older fragmented paradigm: we are at the "brink of ecological and social catastrophe"; we are aware of the "total illusion" of "security through military defense"; we are aware of the terrible consequences of the global economic system that causes "ever increasing disparity between rich and poor"; and we are aware that we need to save humanity "from imminent and total annihilation." All these are caused by the older, dysfunctional world system of autonomous sovereign nation-states and a flawed, class-controlled economic system operating in coordination with this nation-state system. The seventh paragraph of the Preamble again returns to the new paradigm announced in paragraph two:

> Conscious that Humanity is One despite the existence of diverse nations, races, creeds, ideologies and cultures and that the principle of unity in diversity is the basis for a new age when war shall be outlawed and peace prevail; when the earth's total resources shall be equitably used for human welfare; and when basic human rights and responsibilities shall be shared by all without discrimination.

The statement of holism from paragraph two is here spelled out in greater detail. The "diverse nations, races, creeds, ideologies and cultures" of the world no longer mean incommensurable fragmentation, war, and conflict. They are united within this *Constitution* under a "principle of unity in diversity" that is the basis for this "new age" of peace, justice, protection of rights, and assumption of mutual universal responsibilities by the people of Earth. The integrated ability of the Earth Federation government to deal with our climate crisis must be understood in terms of this fundamental paradigm-shift from fragmentation to holism.

It is important to point out that the Preamble expresses its holism (correctly) as the "principle of unity in diversity." The scientific revolution that has placed holism at the center of all processes within the universe understands that a holistic system is qualitatively different from a system of fragmented autonomous parts. In a holistic system the unity in diversity means that the whole functions well because of the parts and the parts function well because of their integration into the whole. The uniqueness of the parts (diversity) is absolutely essential to the proper functioning of the whole. Parts are not, and cannot, be assimilated or absorbed. Throughout the universe, and throughout the ecosystems of the Earth, there are no wholes without diverse interacting parts. It is the same with the holism of the Earth Federation government under the *Earth Constitution*. The whole is systematically designed, as we shall see, to be the function of a diversity of interacting parts. Different cultures, nations, religions, languages, and races must be preserved. In this government, there is simply no whole without these diverse parts. An integrated, and diverse, human community is assured.

There is an analogy with the concept of health, for example, in a human body when all the organs are functioning and integrated into a harmonious whole. Parts and whole working cooperatively together create health in living things, in natural systems, and in the planetary ecosystem. Fragmentation in all these dimensions means death. Similarly, social fragmentation means war and violence, domination and exploitation. The power generated by social holism and world-system holism transforms these negative consequences into a synergistic flourishing of the whole with the harmonious integration of all its parts. This is what social health is and should be—the proper functioning of a genuine human community. The democratic function of the human community as the integration of unity in diversity will reflect the same holism that we seek to preserve for the biosphere of the Earth.

The commonly reproduced diagram of the Earth Federation government reveals the Earth Federation designed as an integrated and comprehensive *system* for preserving and actualizing this holism. The central authority in the world will be the World Parliament, made up of three houses. First is the House of Peoples, elected from 1000 electoral districts worldwide and soliciting the democratic participation of the peoples of Earth in operations of the Earth Federation. The Provisional World Parliament has created the "Global Peoples Assembly" as part of the World Elections Act (WLA 29) to further involve and empower grass roots participation in the Earth Federation government. To effectively prevent climate collapse and restore the environment it will be absolutely necessary to involve local citizens worldwide in the process of conversion to a sustainable way of living.

Second is the House of Nations, with 1, 2, or 3 representatives from each nation depending on population. The concept of "nationhood" will not have the kind of lethal life or death connotation that it currently has among the 193, mostly militarized, sovereign entities today. Under the *Earth Constitution*, it will be easy to create new nations, depending on people's desires, for minorities or other groups who wish a measure of local self-government. Third, and absolutely vital to the holism of the *Earth Constitution*, is the House of Counsellors, 200 representatives from 20 global regions, whose job will be to serve as a source of expertise (including expertise on sustainability, ecology, etc.) for the Parliament and to represent the whole of the Earth and its interests within the World Parliament.

Sessions of the World Parliament may rotate among the five world capitals to be established on five continents. Every continent, nation, and grouping is to be included. The continents of the world are further divided into ten magna-regions (primarily according to population) and twenty administrative regions. Many appointments require a representative from each of these regions, or from each continental division, assuring a continual, and continually variable, diversity filling the high posts of government.

At its first session, each house of the Parliament will elect a panel of five chairpersons, one from each continent, who will rotate annually as chair of that house, the other four serving as vice-chairs each year (Art.5.6.1). Similarly, the Presidium that manages the World Executive branch of the government will consist of five persons, one from each continental division. These five will rotate annually as the President, with the others serving as vice-presidents each year. Similarly, the Executive Cabinet under the Presidium will be composed of twenty to thirty members with at least one from each of the ten administrative magna-regions of the world and with no more than two members from any one nation.

The principle of unity in diversity, reflected in this system of appointments, will also apply to appointments for the World Attorney's General who, with the World Police, will be responsible for enforcement of Earth Federation laws worldwide. The office of the World Attorney's General will again be run by five members, one from each continental division, with the five rotating in the position of World Attorney General every two years. Similarly, the twenty Regional Attorney Generals will be elected from each of the twenty federal administrative regions of the world system, guaranteeing representation from every part of the planet.

Exactly the same structure applies to the office of the World Ombudsmus. The office will be run by five World Ombudsmen, one from each continental division, who will rotate to the office of Principle World Ombudsman every two years. As with the World Attorneys General, there will be twenty world regional offices run by World Advocates under this office, each elected from one of the twenty federal world regions. The fourth office directly responsible to the World Parliament is the World Judiciary, structured within a Collegium of World Judges numbering up to 60 judges, with an equal number elected from each of the ten magna-regions of the world federal administrative districts. Each of the main branches of the Earth Federation, therefore, including the World Parliament, will be structured according to this principle of unity in diversity to ensure that interdependent diversity is the empowering principle behind the holism.

In the understanding of many advanced thinkers concerning the development and maturing of human consciousness, it is axiomatic that (in the words of Ken Wilber): "increasing interior consciousness is accompanied by increasing exterior complexity of the systems housing it" (2007: 57-58). Yet at the same time, higher levels of consciousness are increasingly simple precisely because they are "integral": inclusive and holistic. Something very similar is the case with the sustainability features of the *Earth Constitution*. The sustainability system set up under the *Constitution* has many, complex interlocking features. As a system staffed by human beings from all over the world, it will function in many ways as the brain of the planet: monitoring the complex phenomena of the planet for holistic biospheric balance and regulating to restore and maintain that balance. Such complexity is necessary for sustainability in relation to the very complex biosphere of the Earth and its huge human population. Yet, as integral and holistic, its operation will be amazingly simple, transparent, and efficient.

The *Constitution* provides a holistic governance system for the Earth. It gives us a way to apply our scientific knowledge concerning sustainability, health, human well-being, and spiritual growth and development to our planet as a whole—consistently and comprehensively. As Wilber says of science:

> Rational science here comes to the fore, bringing with it an extraordinary boon for humankind in terms of reduction of suffering and increase in longevity. Counting disease, hunger, illness, and infant mortality, rational science has alleviated more actual human suffering than all the prerational mythic religions combined. That science can be misused is not the issue, its positive gains are staggering and undeniable. (2007: 128)

Scientific knowledge is misused today by two giant constituencies that dominate the economic and political life of humankind: multinational corporations putting profit before people and sovereign nations sponsoring war and militarism. The holistic structure of the *Earth Constitution*, giving representation to the diverse people of Earth and their planetary common good, alone makes possible the application of science for human and planetary welfare. The system set up under the *Constitution* is designed, not only to protect democracy and human rights worldwide, but to make possible the application of science, green technology, and the accumulated wisdom of humankind for the benefit of all persons on Earth

The Integrative Complex is a structural feature of this sustainability system designed to enhance this holistic simplicity. It includes a series of seven agencies set up to serve the functioning of all aspects of the Earth Federation government. It represents the scientific knowledge of humankind now made available for the human and planetary common good. The research, knowledge, assessment, and planning capacities of these agencies will be drawn upon by all the other organs of the Earth Federation, hence "integrating" the knowledge and functioning of the diverse departments and branches of the government. One fundamental way to deal with the series of global crises identified in the Preamble and Article 1 of the *Constitution* is to establish an integrated, diverse, and smoothly functioning world system directed toward addressing these crises in an efficient, scientific, democratic, compassionate, and holistic manner. Along with the rest of the Earth Federation government, it will serve a number of key functions as part of the holistic "brain" of our planet. The *Constitution* provides our planet as a whole, for the first time, with a "brain."

3.2   The Authority for Sustainability in the Earth Constitution

Article 1 of the *Constitution* sets out the "broad functions" of the Earth Federation government. Of the six broad functions listed, number five is "to protect the environment and the ecological fabric of life from all sources of damage, and to control technological innovations whose effects transcend national boundaries, for the purpose of keeping Earth a safe, healthy and happy home for humanity." Hence, sustainability is explicitly established as a central purpose of the Earth Federation. But what of the other five central functions of the Earth Federation? The fact is that these are all deeply interrelated with one another. The first central function is the prevention of war and the securing of disarmament.

We have seen that a sustainable civilization will require fundamental transformation in the current world system across the board, and this will necessarily include the elimination of militarism. The production phase of manufacturing weapons of war creates huge quantities of many biosphere destroying toxic wastes, the transportation and deployment of weapons of war requires vast expenditure of greenhouse gases and the use of fossil fuels, and the use of weapons of war is vastly toxic for the environment. Vietnam and Southeast Asia remain cesspools of the remains of millions of tons of chemicals used on them during the Vietnam War, causing high rates of cancer and birth-defects. The people of southern Iraq and now Afghanistan are experiencing the same high rates of cancer and birth-defects due to the use of depleted uranium weapons in their lands (Sanders 2009, Caldicott 1994). All told, not only is the worldwide war-system toxic to the environment but close to two trillion dollars of the world's wealth are flushed down that toilet annually that absolutely must be used instead for restoring the damaged environment if we are to have a future at all upon this planet.

The same is true of all six broad functions of the Earth Federation. Not only is sustainability explicitly one of the six functions, but the framers of the *Constitution* understood that a decent and sustainable civilization requires an integrated transformation across the board. Consider the six: (1) to prevent war, secure disarmament, and resolve all disputes peacefully, (2) to protect universal human rights including equal opportunities in life, (3) to establish equitable economic and social development for all peoples, (4) to regulate world trade, communications, transportation, currency, standards, use of world resources, and other global and international processes, (5) to protect the ecological fabric of life from all sources of damage, (6) devise solutions to all problems beyond the scope of national governments.

As many of the books by sustainability experts cited above declare, a sustainable world system will require a transformation of the present world system across the board. It will require a new economics, a new equitability and sense of community worldwide, a new democratic spirit worldwide, the elimination of poverty, and the elimination of war. For example, the present massive poverty of 60% of the world's population is deeply unsustainable. Poverty forces people to use wood for cooking fuel and to continually clear new land of forests in the attempt to produce sufficient food for survival. Poverty also pressures people to produce many children who serve as social security as well as labor in their attempt to survive.

A sustainable civilization requires human beings everywhere to see themselves as a community working together for the common good of all and for future generations. This can only happen if egregious poverty has been eliminated, if human rights are protected worldwide, if there are common "world standards" and equitable "use of world resources," and if the Earth Federation is able to deal with all problems beyond the scope of national governments. Sustainability, therefore, requires an integrated world system that deals with all these problems simultaneously. The six broad functions of the Earth Federation government together give us a sustainable world civilization, not number five alone. This is why world government is necessary for sustainability, because there must be a consistent, coordinated, and equitable transformation everywhere on Earth, including disarmament, something than can only be successfully effected through global democratic government.

Article 2 of the *Constitution* states that the Earth Federation shall be non-military, democratic, and based on the sovereignty of the people of Earth. All three of these features are necessary to a sustainable civilization. Militarism, one of the most pervasive, toxic, and unsustainable of all human practices will have to be eliminated, and there is no way that this can happen as long as nation-states remain "sovereign" and see themselves as autonomous territories in competition with rivals and enemies. Democracy , the protection of human rights, and the elimination of poverty will have to be universal, things that are only likely to happen when the people of Earth are recognized as sovereign and united together in a common civilizational project.

Article 4 of the *Constitution* lists 41 "specific powers" granted to the Earth Federation government. If the point I am making here is valid, then it should be clear that all 41 will contribute to sustainability, since this requires a comprehensive transformation of our present world system. For example, "specific power" number 1 is necessary for sustainability: "Prevent wars and armed conflicts among the nations, regions, districts, parts and peoples of Earth." However, despite the cogency of this point, of these 41 specific powers, 16 of them are specifically directed toward protecting the environment and creating a sustainable world system. These are as follows:

4.10 Provide assistance in the event of large scale calamities, including drought, famine, pestilence, flood, earthquake, hurricane, ecological disruptions and other disasters.

4.12 Define standards and promote the worldwide improvement in working conditions, nutrition, health, housing, human settlements, environmental conditions, education, economic security, and other conditions defined under Article 13 of this *World Constitution*.

4.17 Establish and operate world financial, banking, credit and insurance institutions designed to serve human needs; establish, issue and regulate world currency, credit and exchange.

4.18 Plan for and regulate the development, use, conservation and recycling of the natural resources of Earth as the common heritage of Humanity; protect the environment in every way for the benefit of both present and future generations.

4.20 Develop and implement solutions to transnational problems of food supply, agricultural production, soil fertility, soil conservation, pest control, diet, nutrition, drugs and poisons, and the disposal of toxic wastes.

4.21 Develop and implement means to control population growth in relation to the life-support capacities of Earth, and solve problems of population distribution.

4.22 Develop, protect, regulate and conserve the water supplies of Earth; develop, operate and/or coordinate transnational irrigation and other water supply and control projects; assure equitable allocation of trans-national water supplies, and protect against adverse trans-national effects of water or moisture diversion or weather control projects within national boundaries.

4.23 Own, administer and supervise the development and conservation of the oceans and sea-beds of Earth and all resources thereof, and protect from damage.

4.24 Protect from damage, and control and supervise the uses of the atmosphere of Earth.

4.27 Develop, operate and/or coordinate transnational power systems, or networks of small units, integrating into the systems or networks power derived from the sun, wind, water, tides, heat differentials, magnetic forces, and any other source of safe, ecologically sound and continuing energy supply.

4.28 Control the mining, production, transportation and use of fossil sources of energy to the extent necessary to reduce and prevent damages to the environment and the ecology, as well as to prevent conflicts and conserve supplies for sustained use by succeeding generations.

4.29 Exercise exclusive jurisdiction and control over nuclear energy research and testing and nuclear power production, including the right to prohibit any form of testing or production considered hazardous.

4.30 Place under world controls essential natural resources which may be limited or unevenly distributed about the Earth. Find and implement ways to reduce wastes and find ways to minimize disparities when development or production is insufficient to supply everybody with all that may be needed.

4.31 Provide for the examination and assessment of technological innovations which are or may be of supranational consequence, to determine possible hazards or perils to humanity or the environment; institute such controls and regulations of technology as may be found necessary to prevent or correct widespread hazards or perils to human health and welfare.

4.32 Carry out intensive programs to develop safe alternatives to any technology or technological processes which may be hazardous to the environment, the ecological system, or human health and welfare.

4.33 Resolve supranational problems caused by gross disparities in technological development or capability, capital formation, availability of natural resources, educational opportunity, economic opportunity, and wage and price differentials. Assist the processes of technology transfer under conditions which safeguard human welfare and the environment and contribute to minimizing disparities.

This list could have been derived from many of the books on sustainability cited above, from Caldicott's *If You Love this Planet*, Rifkin's *Entropy: Into the Greenhouse World*, Brown's *Eco-Economy*, or Goerner, Dyck, and Lagerroos' *The New Science of Sustainability*. However, there is also a fundamental difference. In such books on sustainability, the steps that need to be taken by the people of Earth are posed as vague abstract principles with very little concrete information about how these steps could actually happen in the face of a system of 193 militarized, competitive nation-states immersed in a globalized capitalism directed toward perpetual growth and the pursuit of profit. In the *Earth Constitution*, these steps are articulated, rather, as "specific powers" granted to a governmental system mandated to establish a sustainable world system. We see, therefore, concretely how it could be done and how it must be done.

Indeed, this list serves as a compendium of 16 essential things that must be consistently done worldwide if we are to live sustainably and protect our planetary biosphere for future generations. All of these 16 are things that must not only be done once by this or that national government (until a coup takes place or a new government is elected). They are not things that must be done here and there around the world (with poor nations complaining that they are not getting help with this from rich nations and rich nations complaining that the poor are taking advantage). They are all essential aspects of a sustainable world system that must be done consistently everywhere and in perpetuity. How could all of these things possibly be systematically done under a system of rival, militarized sovereign nation-states, separated by vast differences of poverty and wealth? The notion is absurd. Sustainability is impossible without a planetary government with the ability to end war, secure disarmament, and to finance, coordinate, and monitor the effectiveness of sustainability measures everywhere on the planet.

3.3   The World Administration Reflects the Constitutional Mandate

Not only is the Earth Federation government empowered and mandated to create democratic sustainable civilization, the administrative structures of its departments, agencies, and branches are designed to fulfill this purpose. Nearly all the departments of the World Administration, for example, will be involved in the maintenance and establishment of sustainable civilization. They are: 7.3.1, Disarmament & War Prevention, 7.3.2 Population, 7.3.3

Food and Agriculture, 7.3.4 Water Supplies and Waterways, 7.3.5 Health and Nutrition, 7.3.6 Education, 7.3.7 Cultural Diversity and the Arts, 7.3.8 Habitat and Settlements, 7.3.9 Environment and Ecology, 7.3.10 World Resources, 7.3.11 Oceans and Seabeds, 7.3.12 Atmosphere and Space, 7.3.13 Energy, 7.3.14 Science and Technology, 7.3.15 Genetic Research & Engineering, 7.3.16 Labor and Income, 7.3.17 Economic & Social Development, 7.3.18 Commerce & Industry, 7.3.19 Transportation and Travel, 7.3.20 Multi-National Corporations, 7.3.21 Communications & Information, 7.3.22 Human Rights, 7.3.23 Distributive Justice, 7.3.24 World Service Corps.

A sustainable civilization will require the participation of all of these departments, which the *Constitution* requires be staffed by highly qualified civil servants with certification by the World Civil Service Administration. We have seen in what way the first of these, disarmament and war prevention, is essential to sustainability, and the second, population, is also universally recognized as essential aspect of sustainability concerns. The third, food and agriculture, is clearly a major concern since at present agricultural lands are disappearing at an alarming rate through unsustainable agricultural practices and the biosphere is becoming polluted through unsustainable use of pesticides and fertilizers. Very similar concerns apply to water supplies and water ways. Health and nutrition will be a vital department for eliminating the many diseases associated with both poverty and climate change.

The sixth department, education, will be absolutely essential to establishing sustainable civilization. Radio, TV, Internet, and schools will all have to offer pervasive education regarding sustainable practices, agriculture, consumption, production, waste disposal, etc. The seventh, cultural diversity and the arts, will also be very important for we saw above that holism requires a real and effective integration of unity with diversity, not an assimilation of diversity into a stultifying sameness. Habitat and settlements will be essential for helping the people of Earth, many of whom are now crowded into vast, unsustainable, polluted cities or massive refugee camps, find sustainable housing and population distribution suitable to their climate and geography. Number nine, environment and ecology, is obviously adapted to the constitutional mandate, as is ten, world resources, eleven, oceans and seabeds, and twelve, atmosphere and space. Number thirteen will also be a vital department, for energy must be sustainable energy derived from solar power, wind, geo-thermal power, tides, and other clean sustainable sources.

The fourteenth department, science and technology, is vital for converting scientific research and technological development from their current focus on war and weapons to sustainable design, conservation, energy, health, etc. The fifteenth, genetic research and engineering, will be vital in regulating, controlling, and prohibiting genetic engineering that threatens the balance of the ecosystem or the health and well-being of creatures. Currently this engineering is done in secret by giant multinational corporations with little oversight, a pure profit motive, and certainly no conception of a planetary common good. The next three departments: labor and income, economic & social development, commerce & industry are clearly all essential to converting the planet to a sustainable economy. Income and social development must create reasonable global equity and "development" cannot any longer mean growth but must mean qualitative transformation of people's lifestyles.

Transportation and travel must create methods of mass transport (by air, water, and land) that no longer use fossil fuels. The department focusing on multi-national corporations must convert these entities from the monsters of unsustainability that many of them currently are to energy friendly and sustainable enterprises. As we have seen, communications and information must be directed at massive multi-media education in sustainability for the people of Earth. Departments of human rights and distributive justice are absolutely necessary if we are to create an interdependent planetary community without the need for militarism, war, terror, or massive security operations.

Finally, the Department for the World Service Corps will be absolutely necessary to restoring the planetary environment as well as activating the planetary economy. The world currently has hundreds of millions of underemployed or unemployed people living in desperate poverty. Like the New Deal under President Roosevelt during the great depression of 1929 in the U.S., the World Service Corps will employ these hundreds of millions of people in cleaning up the environment, proper disposal of waste, establishing solar panels and other environment-friendly technologies, replanting forests, and creating supplies of fresh water for people all over the Earth. This can all be paid for by Earth Federation government created money as we shall see below. It should be clear that every department in the Earth Federation administration will be integrated into the effort to create and maintain a truly sustainable world system for the Earth and for future generations.

3.4   The Integrative Complex Reflects the Constitutional Mandate

We have seen that the Integrative Complex comprises a group of seven agencies that do the research, planning, and monitoring necessary to the functioning of the World Parliament and the other organs of the Earth Federation government. Four of these agencies will be particularly vital to a sustainable world system: the Institute on Governmental Procedures and World Problems, the Agency for Research and Planning, the Agency for Technological and Environmental Assessment, and the World Financial Administration (which I will treat separately in the next subsection). But we have already seen the principles that a successful sustainable world system will involve all agencies of government as well as all nations and peoples on Earth, providing the Earth, for the first time, with its own "brain."

Hence, the World Civil Service Administration will certainly need to tailor its examinations for qualified civil servants to include an understanding of the principles of sustainability and the Commission for Legislative Review will need to understand sustainability deeply in order to advise the Parliament on the cogency, consistency, and possible implications of its legislation directed toward a sustainable world system. Similarly, the World Boundaries and Elections Administration, which is required to conduct a global census every five years, and to "prepare and maintain complete demographic analyses for Earth" will play a vital role in the project of limiting the population of Earth to sustainable proportions. In today's world there is very little systematic monitoring of the world's population and demographics, and speculative estimates of these things vary from one source to another. A sustainable world system will require complete and objective information regarding all categories of the biosphere, one of which is the human population.

The Institute on Governmental Procedures and World Problems is directed to set up systems of education and training concerning world problems and effective governmental procedures for dealing with these problems for all persons in the service of the world government, including those serving in the World Parliament. It will bring in qualified experts from public and private universities, research institutes and activist organizations to activate a process of education, dialogue, and effective action addressing these world problems. The reader will recall that Article 1 of the *Earth Constitution* identified addressing the broad categories of world problems (including the creation of a sustainable planetary civilization) as the primary functions of the Earth Federation government. The Institute is designed to make sure that these problems are addressed throughout the government with the latest scientific information, knowledge, and understanding.

The Institute on Governmental Procedures and World Problems is complemented by the other agencies in the Integrative Complex. The *Constitution* reads: "The functions of the Agency for Research and Planning shall be as follows, but not limited thereto: To serve the World Parliament, the World Executive, the

World Administration, and other organs, departments and agencies of the World Government in any matter requiring research and planning within the competence of the agency. To prepare and maintain a comprehensive inventory of world resources. To prepare comprehensive long-range plans for the development, conservation, recycling and equitable sharing of the resources of Earth for the benefit of all people on Earth, subject to legislative action by the World Parliament. To prepare and maintain a comprehensive list and description of all world problems, including their inter-relationships, impact time projections and proposed solutions, together with bibliographies." This agency, therefore, has the mandate to directly concern itself with *research, planning, and maintenance for a sustainable world system.*

It is important to remind the reader that no such agencies exist in today's world. The U.N. is drastically under-funded and under-staffed for any such comprehensive undertaking such as keeping track of the world's resources and their equitable sharing, recycling, and conservation. The U.N. is also paralyzed by global capitalism and the system of sovereign nation-states that dominate its structure and functioning. A truly sustainable world (and human survival—future generations—depend on our achieving a truly sustainable world) will require agencies like this one and the Institute on World Problems, with the worldwide scope, funding, and expertise to carefully monitor and research the global environmental condition and the effects of human activities on that condition. The same is true for the Agency for Technological and Environmental Assessment.

One of the key factors generating the vast unsustainability of our current world has been technology. Industries have the capacity to alter the environment in drastic ways that were inconceivable only a few decades ago. Logging companies can cut down forests with huge machines that do what not long ago required thousands of men. Coal companies can remove the entire tops of mountains today and sift through the rubble for low grade coal deposits. Oil companies can tap deposits a mile under the ocean and mechanize the extraction of these vast resources of fossil fuels. Genetically engineered organisms and crops have the potential for becoming an invasive species and wiping out the biodiversity of entire lakes, regions, or other ecosystems. The Agency for Technological and Environmental Assessment has the responsibility for carefully monitoring the uses of technology worldwide for their environmental impacts and possible unforeseen consequences. The so-called "environmental impact" studies done today before projects are undertaken are hopelessly inadequate for protecting our planetary biosphere and converting "development" to truly sustainable practices.

The Agency for Technological and Environmental Assessment will maintain a registration and description of all technological innovations, together with impact projections. It will assess the impact of all technologies with respect to the sustainability of the local and planetary ecosystems. It will maintain a global monitoring network in order to accurately measure the effects of technology on the environment. Its central mandate is: "To examine, analyze and assess environmental and ecological problems, in particular the environmental and ecological problems which may result from any intrusions or changes of the environment or ecological relationships which may be caused by technological innovations, processes of resource development, patterns of human settlements, the production of energy, patterns of economic and industrial development, or other man-made intrusions and changes of the environment, or which may result from natural causes." It will be clear to the reader by now that the central mission of the agencies in the Integrative Complex, supplying information and planning for all the organs of the Earth Federation government, including the World Parliament, is to establish and maintain a sustainable civilization for the Earth.

The Integrative Complex will serve as a set coordinating institutions to make the Earth Federation a dynamic, interactive whole. But what we have described so far about the Earth Federation shows that *the entire system of the Federation will be a dynamic*

whole. This can perhaps be best modeled in the geometrical fig-
ure of a torus (a dynamic, donut shaped model). ("In geometry, a
torus (pl. tori) is a surface of revolution generated by revolving a
circle in three-dimensional space about an axis coplanar with the
circle.") All heads of departments of the World Administration
and Integrative Complex will be Ministers, elected to the World
Executive from the World Parliament with specific requirements
of diversity from around the world. This means that the heads
of departments in the government will be returning to their home
constituencies around the world regularly, for they will be representative of electoral districts, world administrative regions, or
nations everywhere on Earth. They will also regularly return to
government as legislators and (those that are heads of agencies or
departments) as administrators.

The dynamism of this model can best be envisioned as a geometrical torus with constantly moving circles bringing people
into the Earth's population, circulating within government with
diverse responsibilities, out to the people, and then in again perpetually. Professional skill and stability within this dynamism will
be provided by the World Civil Service Administration, which
will provide "a Senior Administrator to assist the minister and
supervise the detailed work of the department or agency" (7.2.3).
A dynamic holism is set up with a global holistic movement in
conjunction with an integral professional continuity and on-going
knowledge base.

In the *Constitution*'s description of the Agency for Technological and Environmental Assessment, another feature of the holistic
earth civilization created by the *Constitution*, is again apparent.
For the Earth Federation government will not and cannot be a
global organization apart from, or in conflict with, the community
of the peoples of Earth. Rather, it will have to foster and help establish an Earth community—a real sense of one human reality integrated into one fragile planetary ecosystem (perhaps envisioned
best by this model of a dynamic torus). This agency is required to
"enlist the voluntary or contractual aid and participation of pri-

vate and public universities and colleges, research institutions and other organizations" worldwide in its work of promoting the sustainable uses of technology, assessing the planetary environment with regard to human technological and productive activities, and devising alternatives to harmful uses and activities. Without the help of the planetary non-governmental society, that is without a global sense of a true community working together to create and maintain sustainability, government alone will not be able to do it. The principle of unity in diversity (the holism) that provides the framework and spirit of the *Earth Constitution* dictates that the Earth Federation government enlist the cooperation and energy of all the people of Earth in dealing with all the potentially lethal world problems that it is tasked to address.

3.5 The World's Current Unsustainable Debt-Based Monetary System

To understand the extraordinary power of the stipulations for the World Financial Administrations and their implications for sustainability it is first necessary to look at the current unsustainable, debt-based world monetary system. We have seen that the economists who have worked out the economics of sustainability (such as Herman E. Daly) call for fundamental changes in the way business and production operate. Specifically, Daly (and Rifkin and others) show that economic formulas can no longer treat the economy as a self-contained institution ignoring the finitude of resources, the delicate balance of the planetary ecosystem, or the natural laws of energy and entropy.

Business can no longer operate with calculations of private profit that maximize gain by externalizing the true costs of production to society and the natural environment. The cost calculations must now take into account society, nature, and future generations, which means, as Daly puts it, that the business is no longer independent of values or the human community within which it functions. He writes:

Distribution and scale involve relationships with the poor,
the future, and other species that are fundamentally social
in nature rather than individual. *Homo economicus* as the
self-contained atom of methodological individualism, or as
the pure social being of collectivist theory, are both severe
abstractions. Our concrete experience is that of "persons in
community." We are individual persons, but our very individual identity is defined by the quality of our social relations. Our relations are not just external, they are also internal, that is, the nature of the related entities (ourselves in
this case) changes when relations among them changes. We
are related not only by a nexus of individual willingnesses
to pay for different things, but also by relations of trusteeship for the poor, the future, and other species. The attempt
to abstract from these concrete relations of trusteeship and
reduce everything to a question of individual willingness
to pay is a distortion of our concrete experience as persons
in community, an example of A.N. Whitehead's "fallacy of
misplaced concreteness." (1996: 55)

The global economic meltdown that began in 2008 has led to
a chorus of voices from economists and monetary theorists who
believe that human welfare and the welfare of our planet should
be the central premise of economics. Some of the concepts associated with "democratic socialism" are now being widely discussed by sustainability economists. Whether these thinkers use
the forbidden "S-word," or speak in terms of some form of "cooperative capitalism," or of economics premised on the "public
good," or economics premised on "persons in community," is ir-
relevant to their consensus. There are two central alternatives:
either have money and the economy controlled by huge private
financial monopolies structured around the accumulation of ever-
greater private wealth for themselves (the present world-system)
or place money, some essential forms of production, and the economy democratically in the hands of the people through a public
banking system. The key is not public ownership of the means of
production. The key is in community ownership of money itself:

of the system of credit, debt, and money-creation directed toward human welfare and sustainable development.

Economist Michael Hudson describes the latter alternative in contradistinction to the bailout for the big private banks that was arranged by the Obama administration:

> The alternative is a century and a half old, and emerged out of the ideals of the classical economic doctrines of Adam Smith, David Ricardo, John Stuart Mill, and the last great classical economist, Marx. Their common denominator was to view rent and interest as extractive, not productive. Classical political economy and its successor Progressive Era socialism sought to nationalize the land (or at least to fully tax its rent as the fiscal base). Governments were to create their own credit, not leave this function to wealthy elites via a bank monopoly on credit creation. So today's neoliberalism paints a false picture of what the classical economists envisioned as free markets. They were markets free of economic rent and interest (and taxes to support an aristocracy or oligarchy). Socialism was to free economies from these overhead charges. Today's Obama-Geithner rescue plan is just the reverse.... The Treasury is paying off the gamblers and billionaires by supporting the value of bank loans, investments and derivative gambles, leaving the Treasury in debt. (2009)

The issue is not the abolition of all private property. This idea functions as a red herring promoted by dominant elites to terrorize populations into not examining the present system carefully and honestly. The *Earth Constitution* explicitly affirms the right of private property (Article 12.1 and 12.16). The real issue involves the question as to whether *the economic infrastructure that provides the means for all business and trade* will be democratically owned by the people as a public utility to be used in the service of universal prosperity and sustainability or whether it will be privately owned to be used in the service of private accumulation of wealth for the rich and making a sustainable civilization clearly impossible. The central issues are banking, money creation, and the structure of property laws in general.

This is as much an infrastructure question as are roads and streets. Government normally builds and maintains roads, streets, electrical systems, sewage systems, and other vital infrastructure because these *make possible the free exchange of goods and services* that constitute a healthy economy. You cannot have all streets and roads the private property of individuals or corporations to be used for private interests without throwing the society into chaos. But this is precisely what the dominant monopoly capitalist ethos advocates: the monetary system must be privately owned by giant financial interests and government must raise money through taxes on the people and debt enslavement of the public to these private interests.

In her 2007 book *Web of Debt*, Ellen H. Brown trances the history of the struggle over money creation in the United States (and thus, to a great extent, in the world). She exposes the sleight of hand that led to the creation of the Federal Reserve Bank in 1913, which was and is really a consortium of giant private banks that create money for the people of the United States *as debt* to the private banks. These banks create the money they "lend" to the people of the United States out of nothing and the indebtedness they create enslaves the people who must then pay taxes that are primarily used simply to keep up with the interest on this privately held debt (2007:4). Brown quotes "Sir Josiah Stamp, director of the Bank of England and the second richest man in Britain during the 1920s, who spoke at the University of Texas in 1927:

> *The modern banking system manufactures money out of nothing.* The process is perhaps the most astounding piece of sleight of hand that was ever invented. Banking was conceived in inequity and born in sin. . . . *Bankers own the earth.* Take it away from them but leave them the power to create money, and, with the flick of a pen, they will create enough money to buy it back again. . . . Take this great power away from them and all great fortunes like mine will disappear, for then this would be a better and happier world to live in.

> . . . But, *if you want to continue to be slaves of bankers and pay the cost of your own slavery, then let bankers continue to create money and control credit.* (2007:2)

The overwhelming agreement among economists and monetary theorists who are not indentured as propagandists for the monopoly capitalist system focuses on a socially-owned banking system. As Steven Zarlenga (2002) points out, democracy is effectively gutted when private interests control its financial infrastructure for society loses control over its own destiny. As Brown shows in detail, financial elites who control the money system use their control not to fund the productive economy but to maximize their wealth through the financial manipulation of markets and casino investments directed toward speculative windfalls. And as Hudson says above, both classical economics and its successor progressive era socialism understood that "governments were to create their own credit, not leave this function to wealthy elites via a bank monopoly on credit creation." Zarlenga writes:

> Lack of money to pay for crucial programs is again not a fiscal but a monetary problem caused ultimately by the false idea that government must get money only by taxation or borrowing.... Behind these problems is the fact that the nation is controlled more from behind the scenes by financial institutions than by citizens through elections. *When society loses control over its money system it loses whatever control it might have had over its own destiny.* It can no longer set priorities and the policies for achieving them. It can't solve problems, which then develop into crises and continually mount up.... This book has shown that it is historically self-evident that the best monetary systems have been controlled and monitored through law, by public authority. Leaving money power in private hands has invited, even assured, disastrous results. This is also consistent with the logic of money: since the money system is a creature of law, it rightfully belongs within government, just as the law courts do. (2002: 655-657)

Monetary theorist Richard C. Cook (2008) agrees. The ability of financial elites to speculate in the unproductive economy must be abolished by law and capital markets should be regulated to maximize the productive economy in which real goods and services are produced to serve human needs. All of these thinkers advocate the same basic principle. Money and its democratic creation and regulation should be treated as a "public utility," available to all citizens as part of their social heritage and human rights. Cook itemizes the key points:

- Private creation of credit for speculative purposes should be abolished, and capital markets should be regulated to assure fairness, openness, and freedom from predatory practices.

- Every national government should have the right to spend low cost credit directly into existence for public purposes— including infrastructure, environmental protection, education, and health care—without incurring new debt.

- The physical backing for every currency in existence should be the actual production of national economies.

- National governments should treat credit as a public utility—like clean air, water, or electricity—and should assure its availability to all citizens as their social heritage and as a basic human right. (Ibid.)

Economist J.W. Smith also agrees. The banking system must be "socially owned" and financial speculation by private financial interests must be abolished at the same time that public investment is made in the real productive economy that supplies goods and services to the people. "Due to a *socially-owned* banking system being more powerful than armies...." he writes, "that power is denied a private banking system because their *property rights* are designed for maximum rights to monopolists and minimum rights for all others" (2009: 169-170). Smith affirms that:

> Monopolies claim a large share of the wealth produced,
> waste enormous amounts of resources, capital, and la-
> bor, restrict the efficiencies of an economy, claim unearned
> wealth, and all this doubles the cost of production.... In
> an efficient economy, with *full and equal rights for all*, there
> are no unearned values. Instead of financing unearned
> monopoly-created values, touchable and usable *use values*
> are financed, created, and bought and sold. Both plan-
> ning and financial control are primarily regional and local.
> (Ibid.144-146)

Smith shows that the structure of property laws today actively prevents both social equality and a fully efficient economy. A socially owned banking system not only enables production of the general wealth that substantially eliminates poverty and creates universal prosperity, it democratizes society, placing the power back in the hands of the people's democratically elected government rather than in the hands of financial elites to whom the government is debt ridden and beholden. The key to an economically transformed world order and democratically run community in the service of the common good is clear and simple. Banking must be a public utility in the service of the people and the community, and property laws must be modified so that people gain full rights to the fruits of their labor and creative ingenuity.

However, as we have seen with theorists of sustainability, this fundamental agreement among monetary theorists is marred, however, by a fatal flaw in their thought: many of these theorists remain trapped within the conceptual straitjacket of the sovereign nation-state. They have liberated themselves from the concept of a privately owned monopoly banking system as well as the property laws supporting this system and have shown the path toward a real salvation for humanity from poverty, misery, disease, war, and ecosystem destruction. However, they fail to realize that there can be no such salvation under the planetary war-system of sovereign nations each operating its own central bank on its own behalf. Cook (2008), for example, writes:

- Monetary systems should be controlled by sovereign national governments, not the central banks which mainly serve private finance. The main economic function of the monetary system should be to assure adequate purchasing power to consume an environmentally sustainable and optimal level of production whereby the basic needs of every person in the world community are satisfactorily met.

- Income security, including a basic income guarantee and a national dividend, should be a primary responsibility of national governments in the economic sphere. A right to adequate purchasing power should be part of every national constitution. (www.richardccook.com/articles/)

Cook's excellent and sound monetary principles are undercut by his naïve assumption that each sovereign government of the 193 or so worldwide must have its locally controlled socially owned monetary system. His motivations are clearly in the right place, since socially owned monetary systems will create "an optimum level of production whereby the basic needs of every person in the world community are satisfactorily met." Yet one cannot liberate human beings through sound economic principles coupled with flawed political principles. We have seen the many ways in which the system of sovereign nation-states has devastated democracy, human rights, and peace for more than four centuries, existing, as they do, in a system that is intrinsically a war system.

Exceptions among these monetary thinkers are Ellen Hodgson Brown and J.W. Smith, both of whom acknowledge the need for a global monetary system under some form of an Earth Federation of nations, even though they both see this in the future and do not recognize its immediate necessity and viability. With regard to solutions to the world's financial mess, Brown writes:

> That sort of model has been proposed by an organization called the World Constitution and Parliament Association, which postulates an Earth Federation working for equal

prosperity and well-being for all the Earth's citizens. The global funding body would be authorized not only to advance credit to nations but to issue money directly, on the model of Lincoln's Greenbacks and the IMF's SDRs [special drawing rights]. These funds would then be disbursed as needed for the Common Wealth of Earth. Some such radical overhaul might be possible in the future; but in the meantime, global trade is conducted in many competing currencies, which are vulnerable to speculative attack by pirates prowling in a sea of floating exchange rates. The risk needs to be eliminated. But how? (2007: 440)

Brown ignores that fact that the Earth Federation Movement has *already* designed a universal currency to replace the "many competing currencies" in their "sea of floating exchange rates." She ignores the fact that the founding of the Earth Community is not a "radical overhaul" of the present broken system (which cannot be repaired because it is based on false premises). Rather, the establishment of the Earth community actualizes the very heart of our civilizational project. Its founding on universal principles is the *only possible way* out of today's mess. Smith puts the matter as follows:

The once powerless are getting stronger and they recognize that the imperial centers are getting weaker. Their many alliances and federations are forming a power that will be difficult to challenge and they can serve notice to the historic imperial nations that the UN be restructured into a democratic and moral forum or they will form their own world governing body, effectively a federation of 80% of the world.... World federalist organizations have been working to have a constitution ready for that momentous day. The World Constitution and Parliament Association (WCPA), as does others, has one ready for revision and acceptance by just such an alliance of nations.... With a name picked and a constitution for that governing body in place, the first order of business should be on how best to move forward on world development and alleviation of both poverty and global warming. (2009: 155-156)

Smith sees more clearly that the immediate solution requires global transformation through a "world governing body," although he fails to recognize how unlikely it is that the flawed premises of the U.N. system can be "restructured" into such a body.

While Brown and Smith appear to be far ahead of Hudson, Zarlenga, and Cook in their recognition that a solution to the world's immense economic inequality and instability will ultimately require an Earth Federation, neither of them sees clearly that a solution to today's economic nightmare is *effectively impossible without a real democratic world governing body under the Earth Constitution* having genuine legal authority to operate the monetary and banking systems, reform property laws, alleviate poverty, end war and militarism forever, and deal with climate collapse and create a sustainable civilization. We have seen above some of the ways that all these problems are interconnected and interdependent within the modern world system. They cannot be solved piecemeal by changes within some of the 193 sovereign nations. An effective Earth Federation is a *necessary condition* for a sustainable and pacific world order.

3.6   The World Financial Administration and Sustainable Economics

The World Financial Administration is established in Article 8.7 with the directive to create a Planetary Accounting Office that makes cost/benefit studies and reports on the functioning of all government agencies including their "human, social, environmental, indirect, long-term and other costs and benefits." It is also directed to create a Planetary Banking System and make the transition to a common global currency, valued the same everywhere. Such a stable and reliable currency will be fundamental to both human prosperity and the ability to create a sustainable civilization. Bernard Lietaer, a European central banker who helped design the single European currency (the Euro), in his book *The Future of Money*, writes:

> Your money's value is determined by a global casino of un-precedented proportions: $2 trillion are traded per day in foreign exchange markets, *100 times more than the trading volume of all stocks in the world combined.* Only 2% of these for-eign exchange transactions relate to the "real" economy re-flecting movements of real goods and services in the world, and 98% are purely speculative. This global casino is trig-gering the foreign exchange crisis which shook Mexico in 1994-5, Asia in 1997 and Russia in 1998. (In Brown, 2007: 213)

Sustainability means, we have seen, a world community with the sense that we are all in this together and must work together to preserve and restore our planetary home. A system in which the well-being of people and their ability to operate sustainably is subject to the casino conditions of global financial speculations cannot be allowed to continue. Such a necessary step as the creation of planetary public banking with a single stable earth currency is clearly a near impossibility for the system of milita-rized sovereign nation-states, with some of them relying for their wealth on the huge banks located within their domains, or for the U.N. which is largely a lackey of this present system.

Article 8, sections 7.16 and 17 of the *Earth Constitution* establish the new economic model that will be absolutely essential for a sustainable civilization:

> 7.1.6: Pursuant to specific legislation enacted by the World Parliament, and in conjunction with the Planetary Bank-ing System, to establish and implement the procedures of a Planetary Monetary and Credit System based upon useful productive capacity and performance, both in goods and services. Such a monetary and credit system shall be de-signed for use within the Planetary Banking System for the financing of the activities and projects of the World Govern-ment, and for all other financial purposes approved by the

World Parliament, without requiring the payment of interest on bonds, investments or other claims of financial ownership or debt. 7.1.7: To establish criteria for the extension
of financial credit based upon such considerations as people available to work, usefulness, cost/benefit accounting,
human and social values, environmental health and aesthetics, minimizing disparities, integrity, competent management, appropriate technology, potential production and
performance.

First, these articles establish for the Earth Federation the authority that all governments have—the power to create money.
But, as we have seen, in the case of most governments and the
present world monetary system, this power has been co-opted
by private banking cartels to make most governments and people think that they can only create money as debt, by going ever-
deeper into debt, as is the case with the 16 trillion dollar national
debt owed by the people of the United States. Under the Earth
Federation government, there will be no shortage of money to
create jobs to hire people for restoring the environment nor to pay
for planetary conversion to sustainable energies and technologies.
For the money is to be created "debt-free." Bernard Lietaer, by
contrast, describes our present system:

Greed and competition are not the result of immutable human temperament. . . . Greed and fear of scarcity are in fact
being continuously created and amplified as a direct result
of the kind of money we are using. . . . We can produce
more than enough food to feed everybody, and there is def-
initely enough work for everybody in the world, but there
is clearly not enough money to pay for it all. The scarcity is
in our national currencies. In fact, the job of central banks
is to create and maintain that currency scarcity. The direct
consequence is that we have to fight with one another in
order to survive. (In Brown, 2007:31)

The second major reform is found in section 7.17 quoted
above. The purpose of the planetary banking system will not

be to make money for the rich but to empower people to create businesses, jobs, social projects, and innovations that eliminate poverty, establish peace and harmony, and actualize sustainability. To this end credit will no longer only be available, as today, to those with prior assets that serve as collateral. Credit will now be available based on "people available to work, usefulness, cost/benefit accounting, human and social values, environmental health and aesthetics, minimizing disparities, integrity, competent management, appropriate technology, potential production and performance." Every responsible adult or organization will have access to credit to be paid back at low or no interest rates since the purpose of the credit is a sustainable and prosperous planetary community not the private wealth and power of a few.

Our global monetary system today is 99% composed of privately created debt-money (Brown 2007). Because of this we live in a world of global scarcity and desperation requiring, as we have seen, massive military training for counter-insurgency warfare and massive military interventions by imperial nations designed to protect and promote the present world domination by a tiny corporate and financial elite. The *Earth Constitution* is explicit that money must be created by the Federation as *debt-free money* addressed to the common good and planetary prosperity.

This debt-free, interest-free money is used to promote the prosperity, free trade, and well-being of the people of Earth while protecting the planetary environment. Individuals, corporations, state and local governments may all take advantage of very low cost development loans and lines of credit that are not premised on exploitation of the debtors in the service of private profit. In addition, primary created (debt-free) money will be judiciously spent for global infrastructure needs by the World Parliament. Money and banking are now used in the service of the common good of the people of Earth and in protection of the "ecological fabric of life" on our planet. The rich can no longer exploit the poor through a system of loans and debt that has so far created such misery for the peoples and nations of Earth.

The Earth Federation now coordinates the international actions of demilitarized nation-states through world laws legislated by the World Parliament. Conflicts are settled through the world court system and violators are subject to arrest and prosecution by the World Attorneys General and the World Police. Similarly, transnational corporations are regulated through the democratic legislation of the World Parliament. Their expertise and organizational infrastructures can now be used to promote universal prosperity while protecting the environment.

Three features of the corrupt oligarchy that now dominates the world economy are eliminated from the start. *First*, military Keynesianism (or militarism used to artificially pump up the economies of nations) is eliminated, since under Articles 2 and 17 all militaries worldwide progressively become illegal. *Second*, legal corporate personhood is abolished, which has turned the once beneficial corporations of the world into monstrous, immortal super-humans, who use their billions of dollars and super-human legal rights to dominate the economy of our planet. *Third*, the *Constitution* also removes the ability of these corporate entities to influence politics, judges, and government officials through massive campaign contributions or other forms of monetary influence. Hence, the key steps necessary to founding a truly democratic and prosperous world order take place with the ratification of the *Constitution*: the hold of the militarized oligarchies now dominating the planet is broken along with the hold of their associates, the banking, corporate, and massive financial oligarchies, and the monetary system of the world is placed in the service of the people of Earth. The founding of world democracy under the *Earth Constitution* accomplishes all this from the very beginning.

3.7 Additional Economic Measures Legislated by the Provisional World Parliament

Article 19 of the *Earth Constitution* gives the people of Earth the right and duty to begin Provisional World Government. This has primarily taken the form of people holding sessions of the Provisional World Parliament (PWP), which has been meeting since 1982 and has met in 12 sessions in different cities around the world to date. (The 13th session is scheduled for Lucknow, India, in December 2013). I will discuss below Article 19 and its significance for creating a sustainable world system, for much of the provisional world legislation (in the 52 world legislative acts that the Parliament has passed to date) has been environmental legislation. PWP legislation is not binding on the final World Parliament activated with the ratification of the *Constitution*. However, it serves as a model, a paradigm, and an elaboration of the letter and spirit of the *Constitution*. In this section, I will simply focus on those world legislative acts passed by the Provisional World Parliament that are meant to elaborate the spirit and letter of the *Earth Constitution* concerning financial measures.

These acts include the creation of a World Economic Development Organization (WLA 2), an Earth Federation Funding Corporation (WLA 7) an Earth Financial Credit Corporation (WLA 11), a Provisional Office for World Revenue (WLA 18), a World Patents Act (WLA 21), a World Equity Act (WLA 22), a World Public Utilities Act (WLA 38), and an act for a World Guaranteed Annual Income (WLA 42). Together they are laying the economic foundations for a global market economy based on human rights, promotion of the common good, and a democratic world order that benefits everyone, not just the present 10 percent of humanity who today own 85 percent of all the global wealth (E. Brown 2007: 271). We have seen that the components of sustainability that economists and environmental scientists have unanimously affirmed include reasonable global equity and prosperity. Economic equity and reasonable prosperity are necessary to a sustainable world system. This understanding is built into the *Earth Constitution*.

As early as the first session of the Parliament in 1982, when WLA 2 was passed creating the World Economic Development Organization (WEDO), the Parliament saw through the deception of debt-based money creation. Among the means of funding for WEDO is the directive to develop the financing potential and procedures defined under Article 8, Section 7, paragraphs 4, 5, and 6 of *Earth Constitution to base finance on people's potential productive capacity in both goods and services*, rather than on past savings. The real source of new wealth in the world is the goods and services created by working people (not financial speculation, interest on debt, etc.), and this legislative act recognizes this as the basis for a solid financial system.

From this principle of funding under the *Earth Constitution*, that is, the creation of debt-free fiat money and credit based on the potential of those funded to produce goods and services, follow all the other principles of the Provisional World Parliament that are building the infrastructure for an equitable and just world order. As we have seen, with government-issued debt-free money, the Earth Federation will hire tens of millions of unemployed people in the Third World and elsewhere to restore the environment, re-plant the forests of the Earth, and restore the degraded agricultural lands of the Earth. This massive effort is absolutely necessary if we are to deal effectively with climate collapse and conversion to sustainability. Even though the *Constitution* gives Parliament the right to levy taxes, no such process is necessary within the sound monetary policy formulated by WEDO.

World Legislative Act (WLA) number 7 for an Earth Federation Funding Corporation facilitates the raising of funds under the current system of national currencies to promote the ratification of the *Constitution* and the transition process to an Earth Federation for the planet. WLA 18 creates a Provisional Office of World Revenue with similar intent. Since it is the responsibility of the people of Earth to save their planet from climate collapse, nuclear war, and other impending disasters, the development of provisional world government becomes a very serious project of the PWP. WLA 18 establishes a Provisional Office of World Revenue with the authority to levy taxes from the use of the world commons: land rents, mining surcharges, resource utilization (including hydrocarbon and carbon resources), technology rents, currency transactions, electromagnetic spectrum, etc.

The choice of the use of the world commons as the source of these taxes indicates the on-going concern of the PWP for conversion to sustainability. The current world of some 193 sovereign nations believes that these nations have the right to exploit the global commons for their own purposes and enrichment (the oceans of the world, the minerals lying beneath these oceans, etc.). Even though some international laws supposedly regulate the ability of nations, for example, to hunt endangered whales, in practice there is no meaningful enforcement and the world is steadily diminished because of this anarchic exploitation by multiple nation-states and their private corporations.

WLA 21 establishes a department of World Patents and Intellectual Property rights (IPRs) with the mandate to establish systems of IPRs that do not disenfranchise and marginalize poor nations of the world as is done by the current system favoring the big corporations and the high technology nations. Global equity will require meaningful technology transfer, especially in those technologies (such as solar) that promote sustainable development. Equitable and just IPRs become essential to the equity and the world-wide community spirit that is necessary for a sustainable planet. Hence this act includes as purposes of this Department:

> To devise a system to royalties on patents and property rights that rewards and protects innovators while simultaneously maximizing technology transfer for rapid and sustainable development worldwide. To protect traditional cultures and indigenous peoples from intellectual piracy through unjustifiable use of patent powers. To work closely with the Agency for Technological and Environmental Assessment and other relevant agencies to monitor genetic engineering and other possibly dangerous technologies and regulate permits for development of such technologies for the protection of the people of Earth."

WLA 22 also contributes a major step toward sustainability.

First it defines the value of the Earth Currency in terms of an hour of work and a basket of common, universal commodities. This will ensure a stable, universal Earth Currency so that the people of Earth may establish the reliable and equitable economics that will form an essential basis for sustainable development and ending major disparities of wealth in the world. Second, this act defines the minimum and maximum wages for the Earth Federation. The minimum wage is established in terms of the rights given in Article 13 (to be discussed further below) that ensure to everyone adequate housing, health care, education, food, social security, and wages. The maximum wage, similarly, will be four times the minimum wage, which, in terms of the high value of Earth Currency, will be quite wealthy. Nevertheless, it will eliminate the current vast, unsustainable, flagrant waste of resources and wealth that now characterize the lifestyles of the super-wealthy 10% of our planet's population.

WLA 38, Public Utilities, allocates funding to children's education, disease treatments that now afflict children, the Provisional World Parliament and other purposes deemed by the PWP as urgently needed. WLA 42 establishes the parameters of a guaranteed annual income (GAI) for citizens of the Earth Federation very much in the spirit of WLA 22 (the Equity Act) and Article 13 of the *Earth Constitution* (both of which effectively eliminate involuntary poverty and hunger from the Earth). The act guarantees GAI payments in the case of job loss, illness, or other calamities. Establishing a sustainable world community requires that a world community spirit is also established in which everyone is included and there are no more excluded and marginalized poor as is today the case with some 60% of the world's population.

We see that the Earth Federation will have a common currency valued the same everywhere, ending speculation in currencies and the domination of "hard" over "soft" currencies. It will also institute the principle of "equal pay for equal work," ending the exporting of production to low-wage areas of the world in order to maximize the rate of exploitation and profit. It will encourage

in numerous ways worker investment and cooperative management in the firms within which people often spend their working
lives. It will distribute the work burden among the working population more equitably, and empower people at the grassroots level
worldwide.

In short, it provides a genuinely "New Deal" for the people
of Earth. The tens of millions hired to restore the environment
will have money to exchange in their local economies. In conjunction with interest-free loans or grants for building infrastructure, sanitation systems, education, healthcare, and many private
and public sustainable new enterprises, local economies will "take
off" in that dynamic circulation of money within communities
that economists speak of as economic health. Once the *militarized
nation-state* is removed (today pouring close to two trillion U.S.
dollars per year down the sewer of militarism) along with gigantic corporate and banking institutions dedicated to extracting private profit from localities into foreign banks of the rich, economic
well-being will not be difficult to achieve.

The *Constitution* guarantees everyone on Earth a minimum
wage entirely sufficient to live with dignity and freedom (under
Article 13). It ensures sanitation systems, essential resources, and
educational systems for everyone. It provides every person on
Earth with free health care, free education, and ample insurance
in case of accident or old age. Provisional world legislation enacted by the Provisional World Parliament under the authority of
Article 19 of the *Constitution* provides every person over age 18
with a guaranteed annual income sufficient to eliminate extreme
poverty and starvation from the Earth (WLA 42).

The world order can be fairly easily transformed into one of
planetary peace with justice, reasonable prosperity, and sustainability. The present world-system of scarcity and domination is
a result of the principle inherent in money created as public debt
to private financial elites and on a global system of maximizing
private profit at the expense of the common good of the people
of Earth. Perhaps the most fundamental secret is in "democratic

money": money issued debt-free in the name of the productive capacity of the citizens of Earth to produce goods and services. Everyone on Earth must be included in the effort to convert our planet to a place of peace, security, sustainability, and justice.

These principles cannot work, we have seen, unless we take the principle of "all" seriously and universalize democracy to every person on Earth. A sustainable civilization requires authentic democratic universalism. This universalization process is the fundamental imperative of our time, for it is integral to sustainability. There is a concomitant aspect of our moral obligation today that also requires us to abjure violence, war, and military service altogether. A sustainable civilization can only exist if we reasonably fulfill the deep promise of history for a level of liberty, equality, and community that embraces everyone on earth in a cooperative and productive human endeavor to affirm life rather than death, to affirm future generations rather than selfishly indulging present unsustainable pursuits, and to affirm the holism of the Earth and its creatures, rather than the divisiveness of nationalism, racism, religious fanaticism, or other forms of fragmentation. The World Financial Administration and the process of economic transformation will be integral to this task.

3.8   The World Ombudsmus and Bill of Rights: Articles 11, 12, and 13

The World Ombudsmus, we have seen, is structured as a worldwide agency with offices of World Advocates active in all 20 of the world administrative regions. The Ombudsmus is dedicated to promoting and protecting the human rights of the people of Earth. Human rights are no longer to be an empty ideal as they currently are under the U.N. Universal Declaration of Human Rights in which nation-states (the ones responsible to protect these rights), routinely ignore the rights of both their own citizens and those of other nations in favor of what they deem to be pragmatic, economic, political, or military priorities. The list of "functions and powers" of the Ombudsmus explicitly state that this office will be concerned with the enforcement and implementation of the rights declared in Articles 12 and 13.

Article 12 presents a detailed list of so-called "political" rights and freedoms such as freedom of thought, assembly, information, religion, research, publication, due process of law, etc. We have seen above in what ways the implementation of real democracy for the Earth is linked to the conversion to a sustainable civilization. However, Article 13 called "Directive Principles" described as "other rights for all inhabitants within the Federation of Earth" includes not only the economic rights that we have already discussed but environmental protections and rights as well. It is not, we have seen, just that the *Earth Constitution* recognizes the tremendous need for a sustainable world system but it also recognizes that such a system is a *right* of every person and a *right* essential to future generations. Not only must the World Ombudsmus protect the environmental rights of the people of Earth but among the duties of the Ombudsmus is the following: "To keep on the alert for perils to humanity arising from technological innovations, environmental disruptions and other diverse sources, and to launch initiatives for correction or prevention of such perils."

Key environmental rights are identified in sections 8, 9, 10, 11, 14, and 15 of Article 13: Protection for everyone against the hazards and perils of technological innovations and developments (8). Protection of the natural environment which is the common heritage of humanity against pollution, ecological disruption or damage which could imperil life or lower the quality of life (9). Conservation of those natural resources of Earth which are limited so that present and future generations may continue to enjoy life on the planet Earth (10). Assurance for everyone of adequate housing, of adequate and nutritious food supplies, of safe and adequate water supplies, of pure air with protection of oxygen supplies and the ozone layer, and in general for the continuance of an environment which can sustain healthy living for all (11). Social Security for everyone to relieve the hazards of unemployment,

sickness, old age, family circumstances, disability, catastrophes of
nature, and technological change, and to allow retirement with
sufficient lifetime income for living under conditions of human
dignity during older age (13). Rapid elimination of and prohibitions against technological hazards and man-made environmental disturbances which are found to create dangers to life on Earth
(14). Implementation of intensive programs to discover, develop
and institute safe alternatives and practical substitutions for technologies which must be eliminated and prohibited because of hazards and dangers to life (15).

It should be clear by now why and how the holistic structure
of the Earth Federation and the explicit directives for its departments and agencies are predicated on a fundamental understanding of the dignity and integrity of human beings in relation to the
natural world. For a sustainable civilization is not only a matter
of pragmatic need for survival, not only a matter of survival of a
species, not a mere calculation of the means necessary to achieve
certain ends. Rather, sustainability is an *end in itself* because persons (if the concept of personhood is taken seriously) have rights
to "life, liberty, and security of person," which means the right to
live in a clean and healthy environment with air, water, land, and
biosphere that makes these things possible. These rights (for the
first time in history) are made explicit in the sections from Article
13 quoted above.

The current world system under the U.N. has supplemented
the Universal Declaration with nine other subsequent human
rights declarations that define the agreed parameters of human
rights regarding racial discrimination, civil and political rights,
economic, social and cultural rights, women's rights to be free of
discrimination, rights to be free of torture or degrading treatment,
rights of children, rights of migrant workers, rights against disap-
pearances, and rights of persons with disabilities. Despite the fact
that all of these conventions have monitoring bodies, they have
no enforcement mechanisms other than the voluntary actions of
nation-states (cf. Donnelly 2003). Each nation-state is presumed

responsible for its own citizens (and not those of other states) and each nation is supposed to have autonomy, meaning that other states are not to interfere in its internal affairs. The result is the violation of all ten of these human rights conventions consistently and worldwide.

The *Earth Constitution* is premised on the fact that without enforcement mechanisms human rights are relatively meaningless. It also for the first time in history includes the environmental rights listed above. The result will be that the Earth Federation government is framed, mandated, and organized to protect human rights, including the right to a sustainable environment. As we have seen, all the branches of the Earth Federation government have this mandate: from the World Parliament to the World Judiciary to the World Executive and Administration to the World Police to the World Ombudsmus. The premise is holism: only a holistic world system can give us a democratic, equitable, and sustainable world system. The rights in Article 13 are not just the purview of the World Ombudsmus, therefore. They are principles for every organ of the Earth Federation to follow.

The entire government of the Earth Federation and the worldwide communities that it inspires and empowers, therefore, will necessarily be dedicated to "protection of the natural environment which is the common heritage of humanity against pollution, ecological disruption or damage which could imperil life," "conservation of those natural resources of Earth which are limited so that present and future generations may continue to enjoy life," and "assurance for everyone of adequate housing, of adequate and nutritious food supplies, of safe and adequate water supplies, of pure air with protection of oxygen supplies and the ozone layer, and in general for the continuance of an environment which can sustain healthy living for all." All these quotations from the *Constitution* are statements of sustainability. The office of the World Ombudsmus will be an independent worldwide government agency whose job will be to see that all the other branches of the Earth Federation government comply with these fundamental human rights for a sustainable world system.

3.9 Articles 14 and 16: Safeguards and Exterior Relations

That the *Earth Constitution* is organized around the principle of sustainability is apparent even in these articles that guarantee a certain autonomy to nations within the Earth Federation and deal with any nations or territories that may remain outside the Earth Federation. Under Article 14 nations have the right to determine their own internal political, economic and social systems as "consistent with the guarantees and protections given under this World Constitution to assure civil liberties and human rights and a safe environment for life, and otherwise consistent with the several provisions of this World Constitution." Since a "safe environment" is a human right under the *Constitution*, there would be no need to mention this again here, but the framers of the *Constitution* clearly want there to be no mistake—nation-states have no right to operate without sustainability.

A similar principle is applied to those groups that may wish to live outside the territory of the Earth Federation. They may do so as long as they do not incite violence and as long as their lands are "kept free of acts of environmental or technological damage which seriously affect [the] Earth." Article 16, World Territories and External Relations, defines (once again) the scope of the authority of the Earth Federation government that will assure sustainability for the planet. World territory will include "all oceans and seas having an international or supra-national character, together with the seabeds and resources thereof, beginning at a distance of twenty kilometers offshore, excluding inland seas of traditional national ownership," "vital straits, channels, and canals," and "the atmosphere enveloping Earth, beginning at an elevation of one kilometer above the general surface of the land, excluding the depressions in areas of much variation in elevation." As long as the global commons of the planet are open to the competitive exploitation of private corporations and nation-states, sustainability is impossible. The *Constitution* is very clear that its authority to ensure sustainability includes the entire planet.

3.10 Ratification and Implementation: Article 17

Article 17 presents the criteria for ratification of the *Constitution* and defines the scope, powers, and mandate of the Earth Federation government through three stages of implementation: the first operative stage, the second operative stage and the full operative stage. The wisdom of separating the ratification process into stages allows us to understand the creation of democratic world government as practical and achievable. The first operative stage is activated when only a minimum of 25 nations have ratified, or some equivalent ratification directly by people in world electoral districts, or some combination of the two.

The great seriousness of the *Constitution* toward establishing sustainability is evidenced by many of the responsibilities delegated to the Earth Federation government in its first operative stage. It must create "an Emergency Earth Rescue Administration, concerned with all aspects of climate change and related factors." And it must "expedite the organization and work of an Emergency Earth Rescue Administration, concerned with all aspects of climate change and climate crises." We have already seen the Provisional World Parliament lay the groundwork for this. The Earth Federation government must create "an Integrated Global Energy System, based on environmentally safe sources," and "a World Oceans and Seabeds Administration" to begin protection of the oceans. It must "expedite the new finance, credit and monetary system, to serve human needs," which we have seen is essential to sustainability.

The needs for conversion to a sustainable civilization are absolute and immediate in Article 17. The Earth Federation government must "expedite an integrated global energy system, utilizing solar energy, hydrogen energy, and other safe and sustainable sources of energy," "push forward a global program for agricultural production to achieve maximum sustained yield under conditions which are ecologically sound," "call for and find ways to implement a moratorium on nuclear energy projects until all problems are solved concerning safety, disposal of toxic wastes and the dangers of use or diversion of materials for the production of nuclear weapons," "outlaw and find ways to completely terminate the production of nuclear weapons and all weapons of mass destruction," "push forward programs to assure adequate and non-polluted water supplies and clean air supplies for everybody on Earth," "push forward a global program to conserve and re-cycle the resources of Earth," and "develop an acceptable program to bring population growth under control, especially by raising standards of living."

All of these programs are essential to sustainability and to converting the Earth as rapidly as possible to sustainability. The Earth requires ecologically sound agricultural production, a cleanup and proper disposal of toxic wastes and the elimination of new sources of toxic wastes including nuclear energy. It needs protection for water sources and other vital resources, and it absolutely needs population control. The real work of transition to a sustainable world system cannot happen under the antiquated system of sovereign nation-states (some of which actively encourage national population growth), nor under uncontrolled global corporate capitalism.

Hence, once the initial government is activated, it will have to focus its resources on the planetary environment before permanent collapse occurs. The duties outlined for the second operative stage of Earth Federation (when 50% of the nations or peoples of Earth have ratified) require pushing forward the programs begun in the first operative stage. In the second stage, the polar caps and the continent of Antarctica are declared world territories to be protected as part of the global commons. The second stage also begins the systematic dismantling of the military systems of the

Earth, something we have already seen as vital to achieving a sustainable civilization.

3.11 Provisional World Government and the Work of the Provisional World Parliament: Article 19

We have seen that the *Constitution* empowers the people of Earth to establish Provisional World Government prior to the ratification process that will activate the first operative stage of the Earth Federation government. The framers recognized, first, that all the "world problems" are interrelated and interdependent and, second, that these must be urgently worked on by Provisional World Government, since the nations, corporations, and U.N. are doing nothing adequate or substantial to stop the carnage, let alone restore the environment.

Article 19 calls for "special commissions on each of several of the most urgent world problems" that will prepare an agenda for possible action by the Provisional World Parliament and Provisional World Government. It mandates the PWP to begin work on dealing with these urgent world problems. As resources become available, the preparatory commissions may be reconstituted as Administrative Departments of the Provisional World Government. We are also required to "expedite the new finance, credit and monetary system, to serve human needs."

To date the primary way that the provisional world government has been activated is through sessions of the Provisional World Parliament. The Parliament has taken seriously the need to deal with "urgent world problems" and has passed some 52 world legislative acts dealing with fundamental issues, many of which relate directly to environmental and sustainability matters. These include WLA 3 (creation of an Oceans and Seabeds Authority), creation of the Emergency Earth Rescue Administration (mandated to convert the Earth to sustainability as soon as possible), creation of the World Environment Ministry (for universal environmental protection), and the World Hydrogen System Authority to research and construct clean energy systems based on hydrogen. The "Manifesto" of the fourth and fifth sessions of Parliament declares the oceans and polar caps under the authority of the people of Earth, and WLA 16, the World Hydrocarbon Resource Act, places the hydrocarbon resources of the planet under the authority of the Earth Federation. The World Water Act defines water as belonging to the planet and the right of every person on earth to clean water of reasonable quality and quantity.

In her 1992 book, *Nuclear Madness*, Helen Caldicott presents the terrifying facts about the worldwide nuclear industry, its vast pollution of the planet, its many dangerous accidents and near accidents, its falsification of data, and its monetary investment in continuing this polluting and irrational form of energy production. A sustainable world system will need to abandon the "madness" of nuclear energy. Caldicott predicts that science will never discover a safe way to dispose of wastes or prevent possible disasters. The Provisional World Parliament has understood the urgent need to eliminate not only nuclear weapons, but all nuclear energy along with all forms of bomb and weapons production. None of this can be part of a sustainable civilization.

In the light of these facts, the PWP has passed WLA 1, prohibiting all weapons of mass destruction (perhaps the most urgent of all needs). WLA 13 specifies in detail a range of activities connected with illegal WMDs that are punishable under world law (research, design, transport, investment, deployment, etc.). WLA 33, the Fissile Production Ban, prohibits the production of fissile materials, nuclear weapons, or other explosive devices and outlines the scope and nature of penalties for violation. WLA 34 defines the scope of the ban on nuclear weapons and the obligations and methods for proper disposal. WLA 35, the Nuclear Contamination Act, mandates the World Health Organization (WHO), with sufficient funding from the Earth Federation, to analyze, report on, and review all radiation research in the world and to recommend health-based safety regulations.

World Legislative Acts 39, 40, and 41 form an interrelated unit. WLA 39 prohibits unauthorized destruction of illegal financial instruments. It requires appropriate accounting, disposition, and retirement of illegal securities (all war-making investments and securities) to the World Disarmament Agency office of the Earth Federation Funding Corporation. WLA 40 creates Indemnity Bonds to facilitate rapid divestment from bomb corporation stock, providing an economic incentive for individuals and corporations to convert their assets to non-weapons investments. WLA 41, "Posting the Stock Law," requires stock markets worldwide to recognize and post the laws banning weapons production and investment. It also allows the enforcement system to list the corporations being investigated for possible non-compliance. The three of these acts together might someday be enforceable by the Provisional World Government as it continues to gain in strength and recognition. They will almost certainly be adopted in some version by the World Parliament during the first operative stage of a ratified Earth Federation government.

Finally, with respect to nuclear issues, there is WLA 50, creation of the Nuclear Power Decommission Fund which defines decommissioning parameters for systematic decommissioning of all currently running plants and the parameters for decommissioning in the case of accidents or failure. It also includes guidelines for negotiations for nuclear facilities in countries outside the Earth Federation. It should be clear that the parliament has accomplished a great deal in its 12 sessions to date, responding to the "urgent necessity" to envision, encourage, and implement a viable and sustainable world system.

3.12   Global Education for Sustainability

The transformation of our chaotic world governed by a variety of powerful non-democratic forces (that include military establishments, multinational corporations, mass media public opinion makers, and imperial nation-states) to a sustainable world governed democratically, coherently, and sustainably will require ratification of the *Constitution for the Federation of Earth* with all the economic and administrative innovations we have outlined above. As a peaceful, unified human civilization is established under the *Constitution*, there will also need to be worldwide education for democracy, tolerance, civilizational harmony, mutual respect, and sustainability. The *Constitution* has provided for this worldwide educational effort as well.

Genuine quality education leads to human growth and maturity, and real maturity today is 'planetary maturity' (Martin 2005). Everywhere in the world people need to develop the maturity of mutual tolerance and respect, of interfaith harmony, and of nonviolence through seeing the stupidity of war and violent conflict. Everywhere they need to develop the maturity of identifying with our common humanity rather than with race, ethnicity, gender, nation, cultural group, or other secondary attributes. We saw above that holistic thinker Ken Wilber calls this understanding of the need for spiritual, cultural, and systemic transformation"integral ecology" (2007: 100).

If we are to have a sustainable civilization, people must develop the maturity of finding meaning and fulfillment in the ecstasy of living, in the joy of cultivating the five senses, in the experience of beauty and the sublime, in the happiness that comes from serving others, and in meaningful communication and social interaction with family, community, and species—not in endless consumption, unnecessary wealth, and foolish waste. As psychologist Erich Fromm expressed this, we must find fulfillment in "being" rather than in "having" (1996). Article 13 of the *Constitution* gives as a purpose of the Earth Federation to"assure to each child the right to the full realization of his or her potential." Our highest human potential lies in "being" in these ways, in actualizing our capacity for flourishing with joy, rather than in having, possessing, and endless consumption.

The purpose of life at all levels of human development, after all, is living itself—with joy, happiness, fulfillment, compassion, love, and all-around well-being. In a word: flourishing. The purpose of life is to flourish with joy in a process of self-actualization and self-transcendence that includes community with all others, harmony with our planetary biosphere, and awareness of the universal spiritual Ground of Being. For these reasons a quality planetary educational system is absolutely necessary for a sustainable civilization. Only an Earth Federation government would have the resources and the means to educate the people of Earth regarding all these dimensions of sustainability

The framers of the *Constitution* understood that a quality planetary educational system is a fundamental necessity for a sustainable civilization. In terms of Abraham Maslow's well-known hierarchy of needs, the Earth Federation government can create the conditions in which the physiological, safety, and belonging need are taken care of (through universal healthcare, reasonable prosperity for all, ending militarism, establishing peace, protecting the environment, etc.). You cannot teach moral and spiritual development in any direct way, but you can provide the conditions that make these possible and give pointers as to what constitutes growth in maturity. Our higher development is only possible when our basic needs are first taken care of. Human moral and spiritual development toward Maslow's higher stages of self-esteem, self-actualization, and self-transcendence are made possible for all human beings through the *Earth Constitution* and the Earth Federation. Education can promote this growth toward the ability to find meaning in the fullness of life itself (in self-actualization and self-transcendence) rather than in endless accumulation and meaningless consumption.

Worldwide education will not be limited to classrooms. Today, the internet is transforming communications and human interactions worldwide. The internet can be maintained as an inherently democratic network linking people directly and immediately in information and communications (that are difficult to control or filter in a top-down fashion). Top-down mass media such as newspapers, radio and TV broadcasting, and big-ticket movie-making are being challenged by a much more democratic mode of human interaction spreading rapidly worldwide. The Earth Federation government will participate in the spread of this democratic networking. The top-down media empires have always protected the few and their private interests of wealth and power accumulation as opposed to the common good of the many and the planet that supports us. Democratizing, we have seen, is essential to sustainability. The Internet will make possible a worldwide common education for sustainability and a universal common human consciousness of unity and diversity that is at the heart of sustainability.

We have seen that Article 4.35 gives the Earth Federation the directive to develop a world university system, and the World Administration will include a Department of Education. Such a system will be crucial for the research and the education necessary for converting the productive capacities and social practices of humankind to sustainable forms. Real understanding of the climate crisis has come primarily from the scientists and thinkers of the world, as we have seen (not from capitalists, politicians, or the mass media), and the research necessary to convert to sustainability will require the expertise of scientists, educators, and thinkers. It will require worldwide education for planetary maturity.

We have also seen the emphasis of the *Earth Constitution* upon education in global problems and scientific understanding for agencies and departments of the Earth Federation government. One of the functions of the Institute on Governmental Procedures and World Problems is "to prepare and conduct courses of information, education and training for all personnel in the service of World Government." This will necessarily include education concerning sustainability. The Institute will also "conduct courses and seminars" to educate members of government in "all areas of world problems, particularly for Members of the World Parliament and of the World Executive," as well as all other leaders within the Earth Federation government. To understand world problems scientifically, and to understand them within the context of the cultures and practices of humankind, means to be able to deal with them effectively.

In the world as we find it today, such systematic education is simply not done, except in a very feeble and inadequate way by some U.N. agencies such as UNESCO and its World Philosophical Forum. It needs to be done systematically and worldwide, as only a global government with adequate resources could achieve. Rather, today our global problems—that threaten the very existence of human life on the Earth—are dealt with in a haphazard, unscientific and highly politicized manner by multinational corporate media, politicians with distorting political agendas, national governments with self-interested and imperial agendas, and a global public opinion that is rife with mythologies, rumors, and an inadequate understanding of science.

Following up on this educational mandate provided by the *Constitution*, the Provisional World Parliament (PWP) at its very first session in 1982 passed World Legislative Act 4 for a Graduate School of World Problems and a World University System to be activated under Provisional World Government. The Graduate School functioned under Dr. Terence Amerasinghe until 2003 when it was subsumed under the newly created Institute on World Problems (IOWP) that continues to the present—giving seminars on world problems and leadership training in many countries. In 2004, the PWP passed a world education act that will apply to all educational institutions in the world under Earth Federation authority of receiving Earth Federation funds (World Legislative Act 26).

The act lists a number of Earth Federation agencies and a number of U.N. agencies (to be integrated into the Earth Federation) that will participate in this worldwide educational effort. In the schools of the world that are under the authority of the Earth Federation, students will be involved with a curriculum that is progressively more sophisticated as they develop from early childhood to adulthood. The curriculum will require thoughtful reflection in six areas of global concern: (1) global issues and problems, (2) the *Earth Constitution* itself, (3) the meaningful establishment of unity in diversity, (4) requirements for world peace, (5) the meaning and maintenance of good government, and (6) the establishment of a meaningful quality of life. Whereas the first of these (global issues and problems) means that all students will be studying sustainability, we have seen above that a sustainable civilization will also require widespread thoughtful understanding in the other five categories as well.

Study of the *Earth Constitution* will reveal its deep organizational features promoting sustainability. Study of the principle of unity in diversity will be essential to sustainable civilization—for without an effective realization of that principle a sustainable civilization is not possible. The same is true for the study of the requirements for world peace. We have seen that global militarism is a major contributor to our current unsustainable planetary disorder and that world peace and disarmament are required for sustainability. The study of what constitutes good government will help ensure that the Earth Federation government is functioning optimally and do its job of maintaining sustainability as effectively as possible. Finally, in depth study of the concept of quality of life will reveal that the quality of life is not a function of endless consumption, excessive wealth, or unnecessary waste. Students will see that development in the quality of life can effectively happen within the framework of sustainable civilization. It will include, as mentioned above:"the ecstasy of living, the joy of cultivating the five senses, the experience of beauty and the sublime, the happiness of serving others, and meaningful communication and social interaction with family, community, and species."

Global education, therefore, is not only fundamental to creating a sustainable civilization but the imperative for this education is a central feature of the *Earth Constitution*. In a multiplicity of ways the *Constitution* draws upon scientists, experts, and educators to contribute to a world system dedicated to sustainability. Nominations for the House of Counsellors come from the students and faculty of the universities of the world, with the expectation that the Counsellors will have real expertise in global problems and their solutions that will be available to the other houses in the World Parliament. Delegates in the World Parliament will be educated in addition by the programs of the Institute on World Problems and the Institute on Technical Assessment. Students of the world will study the above described curriculum. A sustainable civilization requires nothing less: democratizing human knowledge and developing a universal common consciousness premised on a deep understanding of how to live a quality life in harmony with our planetary biosphere.

Chapter 4

A Founded World System Based on Clear Principles

4.1 The Significance of a Founded World Order

THE concept of a truly new beginning, of an origin that at the same time raises human life to a higher level of existence, is a principle of utmost importance for the Earth Federation Movement and founding a sustainable civilization. For the founding principles in such a regime are public and there for all to see, shining through the concrete institutions and arrangements of the emergent Earth Federation. Those organizations of world citizens who are attempting to *evolve* the U.N., to affect small, incremental changes in the unjust and violent disorder of things, fail to understand that the compromises inherent in the evolutionary model of human affairs will forever prostitute the ideals one wishes to achieve. Perpetual compromises with immense systems of economic exploitation and imperial sovereign nation-states have led to disaster after disaster for the people of Earth.

The forces that struggle within the U.N. against evolution in the direction of a sustainable, egalitarian, and just world system also struggle within nations and worldwide to perpetuation their system of outrageous wealth, power, domination, and exploitation. They are among the most powerful forces in the world and will forever stall, water down, or pervert attempts to evolve the world system toward sustainability. We are running out of time, and the forces of reaction are trying to stop the transition to sustainability.

In the ratification of the *Earth Constitution* as an integral, completed, and foundational document lies the assent of humanity to true freedom and dignity. An integral legal system, founded on explicit principles of freedom, justice, prosperity, and peace from the very beginning, avoids the prostitution of deeply held ideals involved in all merely evolutionary models by the forces of unfreedom, injustice, unsustainability, poverty, and war. The *Constitution* embodies its highest ideals within an integrated, holistic, legal system that can substantially actualize these ideals in the daily lives of the people of Earth.

The campaign for ratification of an already completed *Constitution* since 1991 indicates, therefore, a disciplined refusal to offer the *Constitution* as a mere draft that can be forever tinkered with by querulous academic pedants and duplicitous forces that would subvert human progress. The campaign for ratification understands that only *a founded system*, only ratification of a completed document of surpassing brilliance (whatever minor flaws might remain), can establish human freedom upon the Earth. The flourishing (and indeed survival) of human life on our planet can only be accomplished through a founding ratification convention bringing the peoples and nations of the world to a truly higher level of human and political existence.

A logical implication of the principle of a *founded* free republic involves the principle of *living systems* that establish freedom, peace, justice, prosperity, and sustainability. We have seen that all these things go together, and any one of them requires the actualization of all together. A widely held assumption today looks at these ideals as reflections of the subjective attitude of people, nations, or economic managers. If the people in government or business are moral, peaceful, and just, it is believed, then this may lead to a world of peace with justice.

This attitude fails to examine the fragmented and distorted institutionalized systems that block morality, peace, sustainability, and justice no matter who makes-up their institutional participants. If economic institutions are flawed and inherently destructive of people and the environment, it matters little whether the captains of banking and industry are moral or immoral. If the system of sovereign nation-states is inherently a war system and a system of power politics, as many thinkers have argued, it matters little who is president or prime minister of various countries.

The Earth Federation Movement understands that freedom, peace, justice, prosperity and sustainability primarily arise from properly designed institutions. If we live under such institutions, democratically and transparently governed, then the flawed human beings who staff these institutions are much more likely to embrace these goods. People who staff today's dysfunctional and unjust economic and nation-state institutions are, for that very reason, more likely to embrace unfreedom, war, injustice, vast poverty, and unsustainability in a world of obscene power and riches for the few. The *Earth Constitution* establishes these ideals through an integrated and universal world legal system. It constitutes a freedom system, a peace system, a justice system, a prosperity system, and a sustainability system for the people of Earth.

Whatever "spiritual" transformations people may undergo, a liberated society will still require, at this stage of history, a well-written democratic constitution for the entire planet specifying limitations on powers, balance of powers, due process, and other institutionalized features designed to prevent totalitarianism and structure a democratic, sustainable system. Such a constitution must embody the ends or the goals of government and economics (since the latter ultimately cannot operate independently of careful, democratically organized planning and oversight). Such a constitution is our only practical hope. As Jawaharlal Nehru, India's Prime Minister, expressed this: "I have no doubt in my mind that World Government must and will come, for there is no other remedy for the world's sickness" (in Habicht 1987: 22).

If the goals written into this constitution involve economics and politics in the service of preserving and enhancing human liberation. If society is directed toward the satisfaction of everyone's individual basic human needs, the development of their potentialities as human beings, and a way of living that is at peace with the Earth and the future, then the *Constitution* itself can be used to ensure that the mechanisms of government and economics remain devoted to these ends. It can be used to see that these goals do not become sidetracked into serving the interests of the few at the expense of the many. Such a society would be a *founded* society, not one that has evolved. It would be a conscious reorganization of society according to liberating principles.

A founded society, based on self-conscious principles, is necessary for human liberation and conversion to a sustainable civilization. A "founded" society is one established according to principles embodied in a founding document. It is entirely different from an "evolved" society in which slow changes blend new principles with older ones with the resulting lack of self-understanding, lack of self-consciousness, and lack of a clear moral and democratic foundation for society. "Respectable" thought, in today's world, insists on just this sort of slow, counterproductive evolution of global society.

Human history has moved to ever greater self-awareness. People are now in a position to create a founded world-society. This is a crucial point. Instead of living in societies that are the product of a kind of blind social evolution from one social form through another, often retaining vestiges of earlier forms of domination (for example, titles of royalty, exclusivistic private property rights, class privilege, racial segregation, etc.), human beings have become capable of founding societies based on principles: ethical, legal, political, economic, and ecological.

A founded society is one in which the founders have moved to the level of a conscious determination of the nature of the social world in which they are to live. We no longer accept the lie asserting "natural laws of economics and society." We now self-consciously realize that human beings choose their economic and social relations. This moves human life, in Kantian language, from the level of social life based on inclinations, to the level of social life based on the free decision to live by principles. Our level of self-awareness has increased since the 18th century, in part through the work of great thinkers, leaders, and visionaries like some of those referenced in this book.

At the dawn of the twenty-first century, we are, for the first time, in a position to self-consciously *found* a just, democratic, peaceful, and enduring planetary society. Today, enough persons manifest planetary consciousness and planetary maturity to a sufficient degree to understand the need to begin a global, founded society. Only a self-conscious society can be free, just, peaceful, and sustainable. Promoters of the system of domination, from universities to mass media, sense this. That is why representatives of the status quo establishment (educational institutions, businesses, corporate mass-media, and governments) work so hard to keep the population in a condition of childlike lack of self-awareness.

The United States, based on its Constitution, is one example of a founded society. The document spells out not only the balance of powers and the framework for the laws of the founded society, but also the rights and freedoms and privileges (as well as duties) of the citizens of that society. With this 18th century development of a founded society, human beings had discovered some elements in the equation of human liberation, for example, checks and balances, guarantee of political rights and liberties, and so on. For its time, this founded society was a great step forward. Today, it serves as a reactionary fetter on the next great step.

We now understand the society that they founded excluded some essential aspects of human liberation such as an economic system predicated on the common good. This is where the United States Constitution fails. It was founded on the principles of formal, political democracy and did not alter the exploitative framework of the economic system of its day. Nor did it challenge the fragmented and unworkable system of sovereign nation-states. The intention was never to create a full democracy enfranchising those referred to by some founding fathers as "the rabble." And the intention was never to transcend the nation-state system but to create a "great nation." A system was created in which the domination of government (by royalty or monarch) was replaced with domination by the rich, a system existing to this day, and continuing to spread its domination across the globe.

Noam Chomsky appeals, in this regard, to Bertrand Russell and John Dewey, "who disagreed on many things but shared a vision" of what a truly human, decent, self-conscious society would be like. For Dewey, Chomsky writes,

> The "ultimate aim" of production is not production of goods, but "of free human beings associated with one another in terms of equality." The goal of education, as Russell put it, is "to give a sense of the value of things other than domination," to help create "wise citizens in a free community" in which both liberty and "individual creativeness" will flourish, and working people will be masters of their fate, not tools of production. Illegitimate structures of coercion must be unraveled. (1996b: 75-76, italics in original)

A founded society is predicated on ethical and democratic goals such as these and understands that institutions infected with "illegitimate structures of coercion" must be abolished. At the dawn of the 21st century, we have developed a degree of self-awareness allowing us to see the illegitimacy of monopoly capitalism and the system of nation-states. Both are based on "structures of coercion." Both are illegitimate. Our praxis at this point in history must be advocacy of the founded society under the *Constitution for the Federation of Earth* and the praxis of delegitimation of the current world order.

Loyalty to the old illegitimate system dehumanizes and demeans us. As Philip Allott expresses this, *"A legal system which does its best to make sense of murder, theft, exploitation, oppression, abuse of power, and injustice, perpetrated by public authorities in the public interest, is a perversion of a legal system"* (1990: xvii, italics added). The legal systems of nation-states claiming sovereignty are illegitimate. People must see that we are all "pilgrim citizens" of the new order. Our human integrity and dignity is related to our commitment to live under a free, just, and peaceful world-order.

This does not entail giving up multi-cultural group identifications or the diversity of voices across the planet. It means giving up the notion that partial groupings can be sovereign, along with the notion that my identification with this group must entail the objectification and dehumanization of other peoples. Revolutionary transition to a liberated society will require the massive delegitimation of the system as it now exists, leading toward the "conscious reorganization of society."

The praxis required at the dawn of the twenty-first century relies on a quantum leap in the creativity and self-aware input of human beings into their destiny (for the sake of our survival as well). Instead of relying on the working out of contradictions in some"dialectical" processes of economic and social transformations (another mythology), humans must act self-consciously on behalf of themselves and future generations. From the works of the great thinkers, and through our own understanding of the global nightmare to which corporate capitalism and the system of autonomous nation-states has brought the world, we are now able to extrapolate the basic parameters of a non-exploitative, truly human, sustainable global society.

These parameters are no mystery. (1) Establish authentic democracy, not the massive manipulation of public opinion by the mass media and the rich. (2) Create the other aspects of authentic democracy such as separation of powers and built-in protection against tyranny or governmental abuses. (3) Assure political

rights: assembly, speech, press, privacy from government surveillance, due process of law, habeas corpus, and so on. The system
must provide these rights for all equally, not only for those who
can afford outrageously expensive lawyers. It must actively promote freedom of the press (not only for those who can afford to
own one). (4) Guarantee a demilitarized world free of the horror
of wars of all kinds. Peace is very much a human right as the International Philosophers for Peace *Document on World Peace* (2001)
affirms. (5) Guarantee economic rights: every one having a right
to health care, a job, social security, a family, security and safety,
adequate housing, and educational opportunities.

(6) Protect diversity: cultural, individual, racial, ethnic, and
religious diversity. (7) Protect the planetary environment so no
business, nation, or group can legally destroy our planet that sustains the fragile life upon it or compromise the precious heritage
of future generations. (8) Guarantee freedom of religion and spirituality (or to assert no religion or spirituality), and establish a cultural, political, and economic environment that enhances (rather
than, as now, destroys) the possibility of spiritual exploration and
growth. Freedom of religion is practically meaningless in the
present environment that systematically destroys most authentic
spirituality and in which human spiritual needs must attempt to
satisfy themselves within a system promoting the very opposite in
the form of greed, selfishness, egoism, hatred of others, fear, and
hopelessness.

(9) Provide a preamble, or statement of the most general principles, which founds the planetary society on the protection, enrichment, fulfillment, and dignity of all human life (and planetary life that we understand today is inseparably connected with
human dignity). (10) Design an economic system that produces
prosperity and sustainable efficiency without destroying any of
the above principles. None of these principles are far fetched or
out of easy reach if we wish to self-consciously found a planetary
society based upon them (see Marchand, 1979: 10-12).

What must be founded here is something without precedent

in history, although it appears in limited, distorted forms in documents like the French *Declaration of the Rights of Man*, the United States *Constitution* and *Declaration of Independence*, the United Nations *Universal Declaration of Human Rights*, and the *Cuban Constitution* of 1976. All of these documents speak of something "universal" to all human beings. Yet they make these universalist claims from within fragmented systems destroying their own possibility of realization. With advanced computers, communications, transportation, and global society, we are now in a position to abolish the fragmentation and realize these universal principles in an effective manner.

For the first time in human history a non-exploitative, non-destructive society can be founded, precisely because we are more fully aware of our historical situation, and aware of our impending doom if we do not transform our ways. The founded society must include the social, economic, and political dimensions of human existence. Human freedom must be assured, and this must include the minimum requirements for being free in terms of food, housing, and social opportunities. Economics, technology, consumption, population reproduction, and waste disposal must all be converted to sustainable formats. An environment allowing development of spirituality must be created so human greed, hatred, and delusion can be overcome and people's lives can turn to simplicity, contentment, and creative fulfillment. The *unity in diversity* of humankind will seem simple and self-evident to future generations, who will revere us as the founding fathers and mothers of a decent world-order.

No society in history has ever been founded squarely on these principles. No society could have been, since these principles can only be implemented at the planetary scale. Our praxis must revolve around a multiplicity of activities making possible the move from here to there, including delegitimation of the present system in every way possible. Sri Aurobindo writes: "The Nation in modern times is practically indestructible—unless it dies from within" (Basu, p. 109).

This is why the concept of "sovereign nations" must be dele-
gitimized. Nation-states must federate under genuine federal
world government, and we *pilgrim citizens* must be loyal to our
planetary citizenship (Falk, 1992). We are pilgrims precisely because, as Guy Marchand writes, "the world laws that the world
citizen has the duty to respect have not yet been enacted" (1979:
13), or have they? There is a worldwide movement of people signing the pledge of allegiance to the *Earth Constitution*. Many thousands are recognizing the *Earth Constitution* as the needed framework for legitimate world law. Now we need a founding ratification convention for the *Constitution for the Federation of Earth*.

4.2   Blueprint for a Sustainable World System

From our detailed account of the *Earth Constitution*, we can see
that it serves as a living blueprint for a sustainable world system.
This is very good news for the majority of people on Earth who are
disenfranchised and marginalized by the current world system, a
system run by capitalist elites and their cousins, the nation-state
political elites. For if we want a decent future for ourselves and
our children, we no longer have to engage exclusively in a politics of negativity, resistance, and struggle. We have something
entirely positive and world transforming that we can affirm and
work on behalf of: the *Constitution for the Federation of Earth*—a
wholly democratic and positive blueprint for a democratic planetary sustainable civilization. We must work to ratify the *Earth
Constitution* in a wholly affirmative effort to *found*, to *establish*, a
planetary civilization based on the universal principles of democracy, equality, community, and sustainability.

We have seen that environmental destruction (like war,
poverty, injustice, and denial of freedom) is a direct consequence
of our present global political and economic system. If companies
have to consider the bottom line in a competitive situation where
they must make a certain margin of profit or go out of business,
then the incentive to *externalize* costs into the air, water, and soil to

the detriment of the planetary ecosystem and future generations is tremendous. Genuine sustainability can only be achieved when the common good and the welfare of future generations are factored into the economic equation. Sustainability means that the resources taken from the Earth are either replaced fully (for example, lumber can be replaced though replanting forests) or used sparingly until ways can be found to substitute artificial resources for essential natural resources (Daly 1996).

We have also seen that the *Earth Constitution* contains dozens of references to "the environment" and the "ecology" of our planet, indicating that a major premise of the emergent Earth Federation is environmental sustainability. The *Constitution* mundializes those natural resources that are vital to the well-being of humanity and that are limited in quantity or non-renewable (Article 4). Hence, they are taken out of the hands of giant corporate monopolies who today exploit them for the private profit of a few at the expense of most of humanity and future generations. The Provisional World Parliament has taken steps to enable this constitutional mandate, for example, by passing the Water Act at its Eighth Session. Multinational corporations have bought up water rights in India and elsewhere and used their "right to private property" to blackmail ordinary citizens who need water (see Shiva, *Water Wars: Privatization, Pollution, and Profit*, 2002).

In his book *When Corporations Rule the World* (1995), former Harvard Business School professor, David Korten, chronicles the devastation of our natural resources as well as the environment by multinational corporations based in the imperial centers of capital. Natural resources are essential for human well-being and need to be carefully conserved for the well-being of all the Earth's citizens as well as future generations. The Provisional World Parliament created enabling legislation for an the World Oceans and Seabeds Authority to supervise the vast riches of the oceans for the welfare of humanity, oceans now being exploited by predatory nation-states, and private corporations without any democratic governmental supervision.

With the vast power placed in human hands by engines, electricity, and specialized machines, the ecosystems of the Earth began to be destroyed at a rate far beyond the ability of nature to heal and repair damages caused by human interference. The technological revolutions of the 18th and 19th centuries continued into the electronic and digital revolutions of the 20th and 21st centuries placing such power in human hands that human activity in its present forms may well destroy the life-support systems of the entire planet and collapse the fabric of life to the point where higher forms of life can no longer survive upon the Earth.

The forests of the world, we have seen, provide the planetary ecosystem with much of the oxygen that supports all aerobic forms of life. They bind carbon dioxide that is exhaled by most living creatures and produced by all forms of combustion. They moderate the climate, provide habitats for most of the vast biodiversity of the Earth, and draw fresh water from the ocean coasts into the interior of continents. Yet the forests of the Earth are disappearing at the rate of an area one half the size of California each year.

In addition to forests, agricultural soils of the Earth are rapidly disappearing. Unsustainable agricultural practices are rapidly depleting top-soils of the planet to the point where vast areas have become unsuitable for agriculture and have been converted to grazing lands. Yet overgrazing worldwide is turning even these areas on every continent into desert wasteland, places that cannot be used to support most life. Runoff from the use of pesticides is poisoning water supplies and ecosystems. Billions of tons of top-soil are lost each year to erosion because of these unsustainable agricultural practices.

Regarding fresh water, the over-pumping of aquifers and overuse of water is dropping water tables worldwide, causing water crises and shortages in many areas of the world. The cities of the world, in addition, are becoming poisoners of the planet's fresh air supplies. Hundreds of millions of gasoline and internal combustion engines and other sources of air pollution spew pollutants into the air. Yet the atmosphere of the Earth is necessary to support all higher forms of life and is at the heart of the ecosystem of our planet.

These cities also produce immense amounts of polluted water, garbage, and trash wastes that are filling and poisoning countrysides, rivers, and oceans worldwide. At the same time, the human population continues to grow at the rate of 80 million new persons per year, every person of whom requires basic resources, fresh water, clean air, and agricultural and forest resources to support them throughout their life-spans, and every one of whom produces waste materials that are returned to the environment (Caldicott 1992; Renner 1996; Daly 1996; Speth 2004).

The principle of Gaia, the idea that the entire Earth (as it has evolved over its 4.6 billion year existence) forms an encompassing ecosystem, is only slowly becoming understood by large numbers of people. This awareness grows as planetary phenomena signaling the alteration of the entire global ecosystem become widely known. Phenomena such as global warming, melting of the polar ice caps, depletion of the ozone layer, collapsing of entire ocean fisheries, rapid extinction of species on a daily basis, increased planetary disasters and super-storms, and possible inversions of global ocean currents and weather patterns are well understood (Lovelock 1991).

Thoughtful human beings today have understood that human life is inseparable from the web of life on Earth. They have understood that we must alter our economic, social, and political practices rapidly to bring human civilization into harmony with the planetary web of life that sustains us. They understand that all development must be sustainable, that it must support human life in the present in ways that do not diminish the life-prospects of future generations. Today, virtually all societies and all nations are living at the expense of future generations, both of humans and other species (Caldicott 1992; Daly 1996; Speth 2004). Actualization of our life-prospects diminishes their life-prospects. At the current rate of destruction, it is even possible that we will reduce their life-prospects to zero.

The *Earth Constitution* and the work of the Provisional World Parliament have been dedicated to addressing these horrific consequences of the present world disorder. The *Constitution* provides a framework integrating economics, resource conservation, technological monitoring and development, education, and legislation. All these aspects of planetary life must be integrated and harmonized if we are to achieve sustainable civilization. The blueprint provided by the *Constitution* also includes the right and duty to begin action for change here and now. The Provisional World Parliament and provisional world government are emerging realities led by citizens of the Earth who understand that we cannot wait for the privileged capitalist and nation-state elites of the world to make grudging concessions. The Parliament is acting now to facilitate the work of the final Earth Federation once the *Constitution* has been ratified by 25 of more nations or electoral districts.

This premise of our global, democratically conceived, well-being is behind the Parliament's passage of the World Hydrogen Energy Authority (WLA #10) to spearhead research and conversion to renewable clean energy for the world, the Hydrocarbon Resource Act (WLA #16) to conserve, regulate on behalf of a clean environment, and utilize democratically the world's remaining hydrocarbon resources, and the Water Act (WLA #30) that recognizes clean water as a right of all persons and takes steps to protect the Earth's diminishing water resources, restore sources of fresh water to the Earth, and democratically apportion these resources to all persons on Earth. We can work positively both toward ratifying the *Earth Constitution* and toward empowering the work of the Provisional World Government.

Recognizing not only that the global environment is threatened but that it is already seriously damaged (as the *Manifesto of the Earth Federation* demonstrates at length), the Provisional World Parliament at its Second Session adopted WLA #6 creating the Emergency Earth Rescue Administration (EERA). The task of the

EERA is to spearhead the gigantic task of restoring the environment of the Earth once the first operative stage of world government under the *Constitution* has been activated. Millions of trees will need to be planted, major initiatives will be needed to restore diminished agricultural lands, and emergency efforts will be required to reclaim sources and conditions for fresh water for the peoples of Earth. Hundreds of millions of currently unemployed persons can be hired by the Earth Federation for these purposes, thereby diminishing global poverty and activating the global economy for the benefit of everyone.

The Parliament also passed WLA #9 creating, within the World Administration of the Earth Federation, a Global Ministry on the Environment to facilitate conversion to sustainability and staff the EERA. Such momentous tasks, absolutely necessary for a decent future for the Earth, can never be accomplished by the fragmented system of nation-states or the U.N. The U.N., which is a mere confederation of sovereign nation-states, has held three global conferences on the destruction of our planetary environment, as we have seen: in Rio de Janeiro, Brazil, in 1992, Johannesburg, South Africa, in 2002, and Copenhagen, Denmark, in 2009. There is common agreement that these were all complete failures to deal with our environmental crises.

The Provisional World Parliament has created a network of practical, pragmatic, and immediately necessary laws and agencies to deal with the immense problems of global environmental restoration and conversion to sustainability. As we have seen, the very first article of the *Earth Constitution* specifies that the fifth broad function of the Earth Federation will be "to protect the environment and the ecological fabric of life from all sources of damage, and to control technological innovations whose effects transcend national boundaries, for the purpose of keeping Earth a safe, healthy and happy home for humanity." Both the *Constitution* and the Parliament are dedicated to creating a world system adequate to this task.

The *Constitution* explicitly requires the government of the

Earth Federation to protect the ecological fabric of life on Earth, that is, to respect the Gaia principle (which is the principle of sustainability) with all its ramifications. Not only does the *Constitution* make this a primary mandate of the Earth Federation, we have seen, but in its second bill of rights (Article 13) makes respect for the Gaia principle a right of the people of Earth themselves and a "directive principle for the world government" to actualize this right. Article 13, numbers 9, 10, and 11, we have seen, read as follows. People have a right to "protection of the natural environment which is the common heritage of humanity against pollution, ecological disruption or damage which could imperil life or lower the quality of life" (9). "Conservation of those natural resources of Earth which are limited so that present and future generations may continue to enjoy life on planet Earth" (10), and "assurance for everyone of adequate housing, of adequate and nutritious food supplies, of safe and adequate water supplies, of pure air with protection of oxygen supplies and the ozone layer, and, in general, for the continuance of an environment which can sustain healthy living for all" (11).

Clearly, here again, the *Constitution* explicitly recognizes the need for human economic, political, and social institutions to conform to the Gaia principle protecting the whole of the planetary environment for future generations. The *Constitution* itself is our blueprint. The key to a sustainable civilization is not only to promote education concerning the principles of natural ecology. This effort alone is insufficient and will ultimately fail unless the anti-ecological institutions of the modern world, described above, are also transformed according to the scientific principles of natural ecology. Our only real and practical option is to ratify the *Constitution for the Federation of Earth*.

For sustainability to happen, the entire human community must be joined together through the dynamic of genuine unity in diversity that constitutes a complementary principle of *social ecology* in human life uniting all people under non-military democratic world government. Only thus can the Gaia principle become a guiding principle for all human political, economic, and social processes. These principles of human democratic ecology are inseparable from the principles of natural ecology. It is necessary to do for humanity what the natural Gaia principle does for nature. We have seen that the *Earth Constitution* does both. It provides global structures that unify all human beings under the principle of *unity in diversity* and the institutions necessary for a sustainable planetary civilization.

Sustainability and structure must be in harmony. The *Constitution for the Federation of the Earth* joins the two together to create a truly ecological and sustainable world order. Our choice is between climate change and system change. Resistance and criticism to the present system, while appropriate, are not enough. Personal conservation and recycling are not nearly enough. We need a concrete, positive blueprint that we can envision, work for, and actualize. The *Earth Constitution* provides the best possible blueprint for system change—for founding a truly sustainable and democratic world civilization.

Appendix A

The Constitution for the Federation of the Earth

Preamble

Realizing that Humanity today has come to a turning point in history and that we are on the threshold of a new world order which promises to usher in an era of peace, prosperity, justice and harmony;

Aware of the interdependence of people, nations and all life;

Aware that man's abuse of science and technology has brought Humanity to the brink of disaster through the production of horrendous weaponry of mass destruction and to the brink of ecological and social catastrophe;

Aware that the traditional concept of security through military defense is a total illusion both for the present and for the future;

Aware of the misery and conflicts caused by ever increasing disparity between rich and poor;

Conscious of our obligation to posterity to save Humanity from imminent and total annihilation;

Conscious that Humanity is One despite the existence of diverse nations, races, creeds, ideologies and cultures and that the principle of unity in diversity is the basis for a new age when war shall be outlawed and peace prevail; when the earth's total resources shall be equitably used for human welfare; and when basic human rights and responsibilities shall be shared by all without discrimination;

Conscious of the inescapable reality that the greatest hope for the survival of life on earth is the establishment of a democratic world government;

We, citizens of the world, hereby resolve to establish a world federation to be governed in accordance with this Constitution for the Federation of Earth.

Article 1   Broad Functions of the Earth Federation

The broad functions of the Federation of Earth shall be:

1.1 To prevent war, secure disarmament, and resolve territorial and other disputes which endanger peace and human rights.

1.2 To protect universal human rights, including life, liberty, security, democracy, and equal opportunities in life.

1.3 To obtain for all people on earth the conditions required for equitable economic and social development and for diminishing social differences.

1.4 To regulate world trade, communications, transportation, currency, standards, use of world resources, and other global and international processes.

1.5 To protect the environment and the ecological fabric of life from all sources of damage, and to control technological innovations whose effects transcend national boundaries, for the purpose of keeping Earth a safe, healthy and happy home for humanity.

1.6 To devise and implement solutions to all problems which are beyond the capacity of national governments, or which are now or may become of global or international concern or consequence.

Article 2 Basic Structure of the Earth Federation

2.1 The Federation of Earth shall be organized as a universal federation, to include all nations and all people, and to encompass all oceans, seas and lands of Earth, inclusive of non-self governing territories, together with the surrounding atmosphere.

2.2 The World Government for the Federation of Earth shall be non-military and shall be democratic in its own structure, with ultimate sovereignty residing in all the people who live on Earth.

2.3 The authority and powers granted to the World Government shall be limited to those defined in this Constitution for the Federation of Earth, applicable to problems and affairs which transcend national boundaries, leaving to national governments jurisdiction over the internal affairs of the respective nations but consistent with the authority of the World Government to protect universal human rights as defined in this World Constitution.

2.4 The basic direct electoral and administrative units of the World Government shall be World Electoral and Administrative Districts. A total of not more than 1000 World Electoral and Administrative Districts shall be defined, and shall be nearly equal in population, within the limits of plus or minus ten percent.

2.5 Contiguous World Electoral and Administrative Districts shall be combined as may be appropriate to compose a total of twenty World Electoral and Administrative Regions for the following purposes, but not limited thereto: for the election or appointment of certain world government officials; for administrative purposes; for composing various organs of the world government as enumerated in Article 4; for the functioning of the Judiciary, the Enforcement System, and the Ombudsmus, as well as for the functioning of any other organ or agency of the World Government.

2.6 The World Electoral and Administrative Regions may be composed of a variable number of World Electoral and Administrative Districts, taking into consideration geographic, cultural, ecological and other factors as well as population.

2.7 Contiguous World Electoral and Administrative Regions shall be grouped together in pairs to compose Magna-Regions.

2.8 The boundaries for World Electoral and Administrative Regions shall not cross the boundaries of the World Electoral and Administrative Districts, and shall be common insofar as feasible for the various administrative departments and for the several organs and agencies of the World Government. Boundaries for the World Electoral and Administrative Districts as well as for the Regions need not conform to existing national boundaries, but shall conform as far as practicable.

2.9 The World Electoral and Administrative Regions shall be grouped to compose at least five Continental Divisions of the Earth, for the election or appointment of certain world government officials, and for certain aspects of the composition and functioning of the several organs and agencies of the World Government as specified hereinafter. The boundaries of Continental Divisions shall not cross existing national boundaries as far as practicable. Continental Divisions may be composed of a variable number of World Electoral and Administrative Regions.

Article 3   Organs of the Earth Federation

The organs of the World Government shall be:

3.1 The World Parliament

3.2 The World Executive

3.3 The World Administration

3.4 The Integrative Complex

3.5 The World Judiciary

3.6 The Enforcement System

3.7 The World Ombudsmus

Article 4 Grant of Specific Powers to the Earth Federation

The powers of the World government to be exercised through its several organs and agencies shall comprise the following:

4.1 Prevent wars and armed conflicts among the nations, regions, districts, parts and peoples of Earth.

4.2 Supervise disarmament and prevent re-armament; prohibit and eliminate the design, testing, manufacture, sale, purchase, use and possession of weapons of mass destruction, and prohibit or regulate all lethal weapons which the World Parliament may decide.

4.3 Prohibit incitement to war, and discrimination against or defamation of conscientious objectors.

4.4 Provide the means for peaceful and just solutions of disputes and conflicts among or between nations, peoples, and/or other components within the Federation of Earth.

4.5 Supervise boundary settlements and conduct plebiscites as needed.

4.6 Define the boundaries for the districts, regions and divisions which are established for electoral, administrative, judicial and other purposes of the World Government.

4.7 Define and regulate procedures for the nomination and election of the members of each House of the World Parliament, and for the nomination, election, appointment and employment of all World Government officials and personnel.

4.8 Codify world laws, including the body of international law developed prior to adoption of the world constitution, but not inconsistent therewith, and which is approved by the World Parliament.

4.9 Establish universal standards for weights, measurements, accounting and records.

4.10 Provide assistance in the event of large scale calamities, including drought, famine, pestilence, flood, earthquake, hurricane, ecological disruptions and other disasters.

4.11 Guarantee and enforce the civil liberties and the basic human rights which are defined in the Bill of Rights for the Citizens of Earth which is made a part of this World Constitution under Article 12.

4.12 Define standards and promote the worldwide improvement in working conditions, nutrition, health, housing, human settlements, environmental conditions, education, economic security, and other conditions defined under Article 13 of this World Constitution.

4.13 Regulate and supervise international transportation, communications, postal services, and migrations of people.

4.14 Regulate and supervise supra-national trade, industry, corporations, businesses, cartels, professional services, labor supply, finances, investments and insurance.

4.15 Secure and supervise the elimination of tariffs and other trade barriers among nations, but with provisions to prevent or minimize hardship for those previously protected by tariffs.

4.16 Raise the revenues and funds, by direct and/or indirect means, which are necessary for the purposes and activities of the World Government.

4.17 Establish and operate world financial, banking, credit and insurance institutions designed to serve human needs; establish, issue and regulate world currency, credit and exchange.

4.18 Plan for and regulate the development, use, conservation and recycling of the natural resources of Earth as the common heritage of Humanity; protect the environment in every way for the benefit of both present and future generations.

4.19 Create and operate a World Economic Development Organization to serve equitably the needs of all nations and people included within the World Federation.

4.20 Develop and implement solutions to transnational problems of food supply, agricultural production, soil fertility, soil conservation, pest control, diet, nutrition, drugs and poisons, and the disposal of toxic wastes.

4.21 Develop and implement means to control population growth in relation to the life-support capacities of Earth, and solve problems of population distribution.

4.22 Develop, protect, regulate and conserve the water supplies of Earth; develop, operate and/or coordinate transnational irrigation and other water supply and control projects; assure equitable allocation of trans-national water supplies, and protect against adverse trans-national effects of water or moisture diversion or weather control projects within national boundaries.

4.23 Own, administer and supervise the development and conservation of the oceans and sea-beds of Earth and all resources thereof, and protect from damage.

4.24 Protect from damage, and control and supervise the uses of the atmosphere of Earth.

4.25 Conduct inter-planetary and cosmic explorations and research; have exclusive jurisdiction over the Moon and over all satellites launched from Earth.

4.26 Establish, operate and/or coordinate global air lines, ocean transport systems, international railways and highways, global communication systems, and means for interplanetary travel and communications; control and administer vital waterways.

4.27 Develop, operate and/or coordinate transnational power systems, or networks of small units, integrating into the systems or networks power derived from the sun, wind, water, tides, heat differentials, magnetic forces, and any other source of safe, ecologically sound and continuing energy supply.

4.28 Control the mining, production, transportation and use of fossil sources of energy to the extent necessary to reduce and prevent damages to the environment and the ecology, as well as to prevent conflicts and conserve supplies for sustained use by succeeding generations.

4.29 Exercise exclusive jurisdiction and control over nuclear energy research and testing and nuclear power production, including the right to prohibit any form of testing or production considered hazardous.

4.30 Place under world controls essential natural resources which may be limited or unevenly distributed about the Earth. Find and implement ways to reduce wastes and find ways to minimize disparities when development or production is insufficient to supply everybody with all that may be needed.

4.31 Provide for the examination and assessment of technological innovations which are or may be of supranational consequence, to determine possible hazards or perils to humanity or the environment; institute such controls and regulations of technology as may be found necessary to prevent or correct widespread hazards or perils to human health and welfare.

4.32 Carry out intensive programs to develop safe alternatives to any technology or technological processes which may be hazardous to the environment, the ecological system, or human health and welfare.

4.33 Resolve supra-national problems caused by gross disparities in technological development or capability, capital formation, availability of natural resources, educational opportunity, economic opportunity, and wage and price differentials. Assist the processes of technology transfer under conditions which safeguard human welfare and the environment and contribute to minimizing disparities.

4.34 Intervene under procedures to be defined by the World Parliament in cases of either intra-state violence and intra-state problems which seriously affect world peace or universal human rights.

4.35 Develop a world university system. Obtain the correction of prejudicialcommunicative materials which cause misunderstandings or conflicts due to differences of race, religion, sex, national origin or affiliation.

4.36 Organize, coordinate and/or administer a voluntary, non-military World Service Corps, to carry out a wide variety of projects designed to serve human welfare.

4.37 Designate as may be found desirable an official world language or official world languages.

4.38 Establish and operate a system of world parks, wild life preserves, natural places, and wilderness areas.

4.39 Define and establish procedures for initiative and referendum by the Citizens of Earth on matters of supra-national legislation not prohibited by this World Constitution.

4.40 Establish such departments, bureaus, commissions, institutes, corporations, administrations, or agencies as may by needed to carry out any and all of the functions and powers of the World Government.

4.41 Serve the needs of humanity in any and all ways which are now, or may prove in the future to be, beyond the capacity of national and local governments.

Article 5   The World Parliament

Section 5.1   Functions and Powers of the World Parliament

5.1.1 To prepare and enact detailed legislation in all areas of authority and jurisdiction granted to the World Government under Article 4 of this World Constitution.

5.1.2 To amend or repeal world laws as may be found necessary or desirable.

5.1.3 To approve, amend or reject the international laws developed prior to the advent of World Government, and to codify and integrate the system of world law and world legislation under the World Government.

5.1.4 To establish such regulations and directions as may be needed, consistent with this world constitution, for the proper functioning of all organs, branches, departments, bureaus, commissions, institutes, agencies or parts of the World Government.

5.1.5 To review, amend and give final approval to each budget for the World Government, as submitted by the World Executive; to devise the specific means for directly raising funds needed to fulfill the budget, including taxes, licenses, fees, globally accounted social and public costs which must be added into the prices for goods and services, loans and credit advances, and any other appropriate means; and to appropriate and allocate funds for all operations and functions of the World Government in accordance with approved budgets, but subject to the right of the Parliament to revise any appropriation not yet spent or contractually committed.

5.1.6 To create, alter, abolish or consolidate the departments, bureaus, commissions, institutes, agencies or other parts of the World Government as may be needed for the best functioning of the several organs of the World Government, subject to the specific provisions of this World Constitution.

5.1.7 To approve the appointments of the heads of all major departments, commissions, offices, agencies and other parts of the several organs of the World Government, except those chosen by electoral or civil service procedures.

5.1.8 To remove from office for cause any member of the World Executive, and any elective or appointive head of any organ, department, office, agency or other part of the World Government, subject to the specific provisions in this World Constitution concerning specific offices.

5.1.9 To define and revise the boundaries of the World Electoral and Administrative Districts, the World Electoral and Administrative Regions and Magna Regions, and the Continental Divisions.

5.1.10 To schedule the implementation of those provisions of the World Constitution which require implementation by stages during the several stages of Provisional World Government, First Operative Stage of World Government, Second Operative Stage of World Government, and Full Operative Stage of World Government, as defined in Articles 17 and 19 of this World Constitution.

5.1.11 To plan and schedule the implementation of those provisions of the World Constitution which may require a period of years to be accomplished.

Section 5.2 Composition of the World Parliament

5.2.1 The World Parliament shall be composed of three houses, designated as follows:

- **The House of Peoples**, to represent the people of Earth directly and equally;

- **The House of Nations**, to represent the nations which are joined together in the Federation of Earth;

- **The House of Counsellors**, with particular functions to represent the highest good and best interests of humanity as a whole.

5.2.2 All members of the World Parliament, regardless of House, shall be designated as Members of the World Parliament.

Section 5.3   The House of Peoples

5.3.1 The House of Peoples shall be composed of the peoples delegates directly elected in proportion to population from the World Electoral and Administrative Districts, as defined in Article 2.4.

5.3.2 Peoples delegates shall be elected by universal adult suffrage, open to all persons of age 18 and above.

5.3.3 One peoples delegate shall be elected from each World Electoral and Administrative District to serve a five year term in the House of Peoples. Peoples delegates may be elected to serve successive terms without limit. Each peoples delegate shall have one vote.

5.3.4 A candidate for election to serve as a peoples delegate must be at least 21 years of age, a resident for at least one year of the electoral district from which the candidate is seeking election, and shall take a pledge of service to humanity.

Section 5.4   The House of Nations

5.4.1 The House of Nations shall be composed of national delegates elected or appointed by procedures to be determined by each national government on the following basis:

5.4.1.1 One national delegate from each nation of at least 100,000 population, but less than 10,000,000 population.

5.4.1.2 Two national delegates from each nation of at least 10,000,000 population, but less than 100,000,000 population.

5.4.1.3 Three national delegates from each nation of 100,000,000 population or more.

5.4.2 Nations of less than 100,000 population may join in groups with other nations for purposes of representation in the House of Nations.

5.4.3 National delegates shall be elected or appointed to serve for terms of five years, and may be elected or appointed to serve successive terms without limit. Each national delegate shall have one vote.

5.4.4 Any person to serve as a national delegate shall be a citizen for at least two years of the nation to be represented, must be at least 21 years of age, and shall take a pledge of service to humanity.

Section 5.5   The House of Counsellors

5.5.1 The House of Counsellors shall be composed of 200 counsellors chosen in equal numbers from nominations submitted from the twenty World Electoral and Administrative Regions, as defined in Article 2.5. and 2.6., ten from each Region.

5.5.2 Nominations for members of the House of Counsellors shall be made by the teachers and students of universities and colleges and of scientific academies and institutes within each world electoral and administrative region. Nominees may be persons who are off campus in any walk of life as well as on campus.

5.5.3 Nominees to the House of Counsellors from each World Electoral and Administrative Region shall, by vote taken among themselves, reduce the number of nominees to no less than two times and no more than three times the number to be elected.

5.5.4 Nominees to serve as members of the House of Counsellors must be at least 25 years of age, and shall take a pledge of service to humanity. There shall be no residence requirement, and a nominee need not be a resident of the region from which nominated or elected.

5.5.5 The members of the House of Counsellors from each region shall be elected by the members of the other two houses of the World Parliament from the particular region.

5.5.6 Counsellors shall be elected to serve terms of ten years. One-half of the members of the House of Counsellors shall be elected every five years. Counsellors may serve successive terms without limit. Each Counsellor shall have one vote.

Section 5.6   Procedures of the World Parliament

5.6.1 Each house of the World Parliament during its first session after general elections shall elect a panel of five chairpersons from among its own members, one from each of five Continental Divisions. The chairpersons shall rotate annually so that each will serve for one year as chief presiding officer, while the other four serve as vice-chairpersons.

5.6.2 The panels of Chairpersons from each House shall meet together, as needed, for the purpose of coordinating the work of the Houses of the World Parliament, both severally and jointly.

5.6.3 Any legislative measure or action may be initiated in either House of Peoples or House of Nations or both concurrently, and shall become effective when passed by a simple majority vote of both the House of Peoples and of the House of Nations, except in those cases where an absolute majority vote or other voting majority is specified in this World Constitution.

5.6.4 In case of deadlock on a measure initiated in either the House of Peoples or House of Nations, the measure shall then automatically go to the House of Counsellors for decision by simple majority vote of the House of Counsellors, except in the cases where other majority vote is required in this World Constitution. Any measure may be referred for decision to the House of Counsellors by a concurrent vote of the other two houses.

5.6.5 The House of Counsellors may initiate any legislative measure, which shall then be submitted to the other two houses and must be passed by simple majority vote of both the House of Peoples and House of Nations to become effective, unless other voting majority is required by some provision of this World Constitution.

5.6.6 The House of Counsellors may introduce an opinion or resolution on any measure pending before either of the other two houses; either of the other houses may request the opinion of the House of Counsellors before acting upon a measure.

5.6.7 Each house of the World Parliament shall adopt its own detailed rules of procedure, which shall by consistent with the procedures set forth in this World Constitution, and which shall be designed to facilitate coordinated functioning of the three houses.

5.6.8 Approval of appointments by the World Parliament or any house thereof shall require simple majority votes, while removals for cause shall require absolute majority votes.

5.6.9 After the full operative stage of World Government is declared, general elections for members of the World Parliament to the House of Peoples shall be held every five years. The first general elections shall be held within the first two years following the declaration of the full operative stage of World Government.

5.6.10 Until the full operative stage of World Government is declared, elections for members of the World Parliament to the House of Peoples may be conducted whenever feasible in relation to the campaign for ratification of this World Constitution.

5.6.11 Regular sessions of the House of Peoples and House of Nations of the World Parliament shall convene on the second Monday of January of each and every year.

5.6.12 Each nation, according to its own procedures, shall appoint or elect members of the World Parliament to the House of Nations at least thirty days prior to the date for convening the World Parliament in January.

5.6.13 The House of Peoples together with the House of Nations shall elect the members of the World Parliament to the House of Counsellors during the month of January after the general elections. For its first session after general elections, the House of Counsellors shall convene on the second Monday of March, and thereafter concurrently with the other two houses.

5.6.14 Bi-elections to fill vacancies shall be held within three months from occurrence of the vacancy or vacancies.

5.6.15 The World Parliament shall remain in session for a minimum of nine months of each year. One or two breaks may be taken during each year, at times and for durations to be decided by simple majority vote of the House of Peoples and House of Nations sitting jointly.

5.6.16 Annual salaries for members of the World Parliament of all three houses shall be the same, except for those who serve also as members of the Presidium and of the Executive Cabinet.

5.6.17 Salary schedules for members of the World Parliament and for members of the Presidium and of the Executive Cabinet shall be determined by the World Parliament.

Article 6   The World Executive

Section 6.1   Functions and Powers of the World Executive

6.1.1 To implement the basic system of world law as defined in the World Constitution and in the codified system of world law after approval by the World Parliament.

6.1.2 To implement legislation enacted by the World Parliament.

6.1.3 To propose and recommend legislation for enactment by the World Parliament.

6.1.4 To convene the World Parliament in special sessions when necessary.

6.1.5 To supervise the World Administration and the Integrative Complex and all of the departments, bureaus, offices, institutes and agencies thereof.

6.1.6 To nominate, select and remove the heads of various organs, branches, departments, bureaus, offices, commissions, institutes, agencies and other parts of the World Government, in accordance with the provisions of this World Constitution and as specified in measures enacted by the World Parliament.

6.1.7 To prepare and submit annually to the World Parliament a comprehensive budget for the operations of the World Government, and to prepare and submit periodically budget projections over periods of several years.

6.1.8 To define and propose priorities for world legislation and budgetary allocations.

6.1.9 To be held accountable to the World Parliament for the expenditures of appropriations made by the World Parliament in accordance with approved and longer term budgets, subject to revisions approved by the World Parliament.

Section 6.2   Composition of the World Executive

The World Executive shall consist of a Presidium of five members, and of an Executive Cabinet of from twenty to thirty members, all of whom shall be members of the World Parliament.

Section 6.3   The Presidium

6.3.1 The Presidium shall be composed of five members, one to be designated as President and the other four to be designated as Vice Presidents. Each member of the Presidium shall be from a different Continental Division.

6.3.2 The Presidency of the Presidium shall rotate each year, with each member in turn to serve as President, while the other four serve as Vice Presidents. The order of rotation shall be decided by the Presidium.

6.3.3 The decisions of the Presidium shall be taken collectively, on the basis of majority decisions.

6.3.4 Each member of the Presidium shall be a member of the World Parliament, either elected to the House of Peoples or to the House of Counsellors, or appointed or elected to the House of Nations.

6.3.5 Nominations for the Presidium shall be made by the House of Counsellors. The number of nominees shall be from two to three times the number to be elected. No more than one-third of the nominees shall be from the House of Counsellors or from the House of Nations, and nominees must be included from all Continental Divisions.

6.3.6 From among the nominees submitted by the House of Counsellors, the Presidium shall be elected by vote of the combined membership of all three houses of the World Parliament in joint session. A plurality vote equal to at least 40 percent of the total membership of the World Parliament shall be required for the election of each member to the Presidium, with successive elimination votes taken as necessary until the required plurality is achieved.

6.3.7 Members of the Presidium may be removed for cause, either individually or collectively, by an absolute majority vote of the combined membership of the three houses of the World Parliament in joint session.

6.3.8 The term of office for the Presidium shall be five years and shall run concurrently with the terms of office for the members as Members of the World Parliament, except that at the end of each five year period, the Presidium members in office shall continue to serve until the new Presidium for the succeeding term is elected. Membership in the Presidium shall be limited to two consecutive terms.

Section 6.4 The Executive Cabinet

6.4.1 The Executive Cabinet shall be composed of from twenty to thirty members, with at least one member from each of the ten World Electoral and Administrative Magna Regions of the world.

6.4.2 All members of the Executive Cabinet shall be Members of the World Parliament.

6.4.3 There shall be no more than two members of the Executive Cabinet from any single nation of the World Federation. There may be only one member of the Executive Cabinet from a nation from which a Member of the World Parliament is serving as a member of the Presidium.

6.4.4 Each member of the Executive Cabinet shall serve as the head of a department or agency of the World Administration or Integrative Complex, and in this capacity shall be designated as Minister of the particular department or agency.

6.4.5 Nominations for members of the Executive Cabinet shall be made by the Presidium, taking into consideration the various functions which Executive Cabinets members are to perform. The Presidium shall nominate no more than two times the number to be elected.

6.4.6 The Executive Cabinet shall be elected by simple majority vote of the combined membership of all three houses of the World Parliament in joint session.

6.4.7 Members of the Executive Cabinet either individually or collectively may be removed for cause by an absolute majority vote of the combined membership of all three houses of the World Parliament sitting in joint session.

6.4.8 The term of office in the Executive Cabinet shall be five years, and shall run concurrently with the terms of office for the members as Members of the World Parliament, except that at the end of each five year period, the Cabinet members in office shall continue to serve until the new Executive Cabinet for the succeeding term is elected. Membership in the Executive Cabinet shall be limited to three consecutive terms, regardless of change in ministerial position.

Section 6.5   Procedures of the World Executive

6.5.1 The Presidium shall assign the ministerial positions among the Cabinet members to head the several administrative departments and major agencies of the Administration and of the Integrative Complex. Each Vice President may also serve as a Minister to head an administrative department, but not the President. Ministerial positions may be changed at the discretion of the Presidium. A Cabinet member or Vice President may hold more than one ministerial post, but no more than three, providing that no Cabinet member is without a Ministerial post.

6.5.2 The Presidium, in consultation with the Executive Cabinet, shall prepare and present to the World Parliament near the beginning of each year a proposed program of world legislation. The Presidium may propose other legislation during the year.

6.5.3 The Presidium, in consultation with the Executive Cabinet, and in consultation with the World Financial Administration, (see Article 8, Sec. 7.1.9.) shall be responsible for preparing and submitting to the World Parliament the proposed annual budget, and budgetary projections over periods of years.

6.5.4 Each Cabinet Member and Vice President as Minister of a particular department or agency shall prepare an annual report for the particular department or agency, to be submitted both to the Presidium and to the World Parliament.

6.5.5 The members of the Presidium and of the Executive Cabinet at all times shall be responsible both individually and collectively to the World Parliament.

6.5.6 Vacancies occurring at any time in the World Executive shall be filled within sixty days by nomination and election in the same manner as specified for filling the offices originally.

Section 6.6   Limitations on the World Executive

6.6.1 The World Executive shall not at any time alter, suspend, abridge, infringe or otherwise violate any provision of this World Constitution or any legislation or world law enacted or approved by the World Parliament in accordance with the provisions of this World Constitution.

6.6.2 The World Executive shall not have veto power over any legislation passed by the World Parliament.

6.6.3 The World Executive may not dissolve the World Parliament or any House of the World Parliament.

6.6.4 The World Executive may not act contrary to decisions of the World Courts.

6.6.5 The World Executive shall be bound to faithfully execute all legislation passed by the World Parliament in accordance with the provisions of this World Constitution, and may not impound or refuse to spend funds appropriated by the World Parliament, nor spend more funds than are appropriated by the World Parliament.

6.6.6 The World Executive may not transcend or contradict the decisions or controls of the World Parliament, the World Judiciary or the Provisions of this World Constitution by any device of executive order or executive privilege or emergency declaration or decree.

Article 7   The World Administration

Section 7.1   Functions of the World Administration

7.1.1 The World Administration shall be organized to carry out the detailed and continuous administration and implementation of world legislation and world law.

7.1.2 The World Administration shall be under the direction of the World Executive, and shall at all times be responsible to the World Executive.

7.1.3 The World Administration shall be organized so as to give professional continuity to the work of administration and implementation.

Section 7.2   Structure and Procedures of the World Administration

7.2.1 The World Administration shall be composed of professionally organized departments and other agencies in all areas of activity requiring continuity of administration and implementation by the World Government.

7.2.2 Each Department or major agency of the World Administration shall be headed by a Minister who shall be either a member of the Executive Cabinet or a Vice President of the Presidium.

7.2.3 Each Department or major agency of the World Administration shall have as chief of staff a Senior Administrator, who shall assist the Minister and supervise the detailed work of the Department or agency.

7.2.4 Each Senior Administrator shall be nominated by the Minister of the particular Department or agency from among persons in the senior lists of the World Civil Service Administration, as soon as senior lists have been established by the World Civil Service Administration, and shall be confirmed by the Presidium. Temporary qualified appointments shall be made by the Ministers, with confirmation by the Presidium, pending establishment of the senior lists.

7.2.5 There shall be a Secretary General of the World Administration, who shall be nominated by the Presidium and confirmed by absolute majority vote of the entire Executive Cabinet.

7.2.6 The functions and responsibilities of the Secretary General of the World Administration shall be to assist in coordinating the work of the Senior Administrators of the several Departments and agencies of the World Administration. The Secretary General shall at all times be subject to the direction of the Presidium, and shall be directly responsible to the Presidium.

7.2.7 The employment of any Senior Administrator and of the Secretary General may be terminated for cause by absolute majority vote of both the Executive Cabinet and Presidium combined, but not contrary to civil service rules which protect tenure on grounds of competence.

7.2.8 Each Minister of a Department or agency of the World Administration, being also a Member of the World Parliament, shall provide continuous liaison between the particular Department or agency and the World Parliament, shall respond at any time to any questions or requests for information from the Parliament, including committees of any House of the World Parliament.

7.2.9 The Presidium, in cooperation with the particular Ministers in each case, shall be responsible for the original organization of each of the Departments and major agencies of the World Administration.

7.2.10 The assignment of legislative measures, constitutional provisions and areas of world law to particular Departments and agencies for administration and implementation shall be done by the Presidium in consultation with the Executive Cabinet and Secretary General, unless specifically provided in legislation passed by the World Parliament.

7.2.11 The Presidium, in consultation with the Executive Cabinet, may propose the creation of other departments and agencies to have ministerial status; and may propose the alteration, combination or termination of existing Departments and agencies of ministerial status as may seem necessary or desirable. Any such creation, alteration, combination or termination shall require a simple majority vote of approval of the three houses of the World Parliament in joint session.

7.2.12 The World Parliament by absolute majority vote of the three houses in joint session may specify the creation of new departments or agencies of ministerial status in the World Administration, or may direct the World Executive to alter, combine, or terminate existing departments or agencies of ministerial status.

7.2.13 The Presidium and the World Executive may not create, establish or maintain any administrative or executive department or agency for the purpose of circumventing control by the World Parliament.

Section 7.3   Departments of the World Administration

7.3.1 Disarmament & War Prevention.

7.3.2 Population.

7.3.3 Food and Agriculture.

7.3.4 Water Supplies and Waterways.

7.3.5 Health and Nutrition.

7.3.6 Education.

7.3.7 Cultural Diversity and the Arts.

7.3.8 Habitat and Settlements.

7.3.9 Environment and Ecology.

7.3.10 World Resources.

7.3.11 Oceans and Seabeds.

7.3.12 Atmosphere and Space.

7.3.13 Energy.

7.3.14 Science and Technology.

7.3.15 Genetic Research & Engineering.

7.3.16 Labor and Income.

7.3.17 Economic & Social Development.

7.3.18 Commerce & Industry

7.3.19 Transportation and Travel.

7.3.20 Multi-National Corporations.

7.3.21 Communications & Information.

7.3.22 Human Rights.

7.3.23 Distributive Justice.

7.3.24 World Service Corps.

7.3.25 World Territories, Capitals & Parks.

7.3.26 Exterior Relations.

7.3.27 Democratic Procedures.

7.3.28 Revenue.

Article 8   The Integrative Complex

Section 8.1   Definition

8.1.1 Certain administrative, research, planning and facilitative agencies of the World Government which are particularly essential for the satisfactory functioning of all or most aspects of the World Government, shall be designated as the Integrative Complex. The Integrative Complex shall include the agencies listed under this section, with the proviso that other such agencies may be added upon recommendation of the Presidium followed by decision of the World Parliament.

8.1.1.1 The World Civil Service Administration.

8.1.1.2 The World Boundaries and Elections Administration.

8.1.1.3 The Institute on Governmental Procedures and World Problems.

8.1.1.4 The Agency for Research and Planning.

8.1.1.5 The Agency for Technological and Environmental Assessment.

8.1.1.6 The World Financial Administration.

8.1.1.7 Commission for Legislative Review.

8.1.2 Each agency of the Integrative Complex shall be headed by a Cabinet Minister and a Senior Administrator, or by a Vice President and a Senior Administrator, together with a Commission as provided hereunder. The rules of procedure for each agency shall be decided by majority decision of the Commission members together with the Administrator and the Minister or Vice President.

8.1.3 The World Parliament may at any time define further the responsibilities, functioning and organization of the several agencies of the Integrative Complex, consistent with the provisions of Article 8 and other provisions of the World Constitution.

8.1.4 Each agency of the Integrative Complex shall make an annual report to the World Parliament and to the Presidium.

Section 8.2   The World Civil Service Administration

8.2.1 The functions of the World Civil Service Administration shall be the following, but not limited thereto:

   8.2.1.1 To formulate and define standards, qualifications, tests, examinations and salary scales for the personnel of all organs, departments, bureaus, offices, commissions and agencies of the World Government, in conformity with the provisions of this World Constitution and requiring approval by the Presidium and Executive Cabinet, subject to review and approval by the World Parliament.

   8.2.1.2 To establish rosters or lists of competent personnel for all categories of personnel to be appointed or employed in the service of the World Government.

   8.2.1.3 To select and employ upon request by any government organ, department, bureau, office, institute, commission, agency or authorized official, such competent personnel as may be needed and authorized, except for those positions which are made elective or appointive under provisions of the World Constitution or by specific legislation of the World Parliament.

8.2.2 The World Civil Service Administration shall be headed by a ten member commission in addition to the Cabinet Minister or Vice President and Senior Administrator. The Commission shall be composed of one commissioner from each of ten World Electoral and Administrative Magna-Regions. The persons to serve as Commissioners shall be nominated by the House of Counsellors and then appointed by the Presidium for five year terms. Commissioners may serve consecutive terms.

Section 8.3   The World Boundaries and Elections Administration

8.3.1 The functions of the World Boundaries and Elections Administration shall include the following, but not limited thereto:

   8.3.1.1 To define the boundaries for the basic World Electoral and Administrative Districts, the World Electoral and Administrative Regions and Magna-Regions, and the Continental Divisions, for submission to the World Parliament for approval by legislative action.

8.3.1.2 To make periodic adjustments every ten or five years, as needed, of the boundaries for the World Electoral and Administrative Districts, the World Electoral and Administrative Regions and Magna-Regions, and of the Continental Divisions, subject to approval by the World Parliament.

8.3.1.3 To define the detailed procedures for the nomination and election of Members of the World Parliament to the House of Peoples and to the House of Counsellors, subject to approval by the World Parliament.

8.3.1.4 To conduct the elections for Members of the World Parliament to the House of Peoples and to the House of Counsellors.

8.3.1.5 Before each World Parliamentary Election, to prepare Voters' Information Booklets which shall summarize major current public issues, and shall list each candidate for elective office together with standard information about each candidate, and give space for each candidate to state his or her views on the defined major issues as well as on any other major issue of choice; to include information on any initiatives or referendums which are to be voted upon; to distribute the Voters' Information Booklets for each World Electoral District, or suitable group of Districts; and to obtain the advice of the Institute on Governmental Procedures and World Problems, the Agency for Research and Planning, and the Agency for Technological and Environmental Assessment in preparing the booklets.

8.3.1.6 To define the rules for world political parties, subject to approval by the World Parliament, and subject to review and recommendations of the World Ombudsmus.

8.3.1.7 To define the detailed procedures for legislative initiative and referendum by the Citizens of Earth, and to conduct voting on supra- national or global initiatives and referendums in conjuction with world parliamentary elections.

8.3.1.8 To conduct plebiscites when requested by other Organs of the World Government, and to make recommendations for the settlement of boundary disputes.

8.3.1.9 To conduct a global census every five years, and to prepare and maintain complete demographic analyses for Earth.

8.3.2 The World Boundaries and Elections Administration shall be headed by a ten member commission in addition to the Senior Administrator and the Cabinet Minister or Vice President. The commission shall be composed of one commissioner each from ten World Electoral and Administrative Magna-Regions. The persons to serve as commissioners shall be nominated by the House of Counsellors and then appointed by the World Presidium for five year terms. Commissioners may serve consecutive terms.

Section 8.4   Institute on Governmental Procedures and World Problems

8.4.1 The functions of the World Boundaries and Elections Administration shall include the following, but not limited thereto:

8.4.1.1 To prepare and conduct courses of information, education and training for all personnel in the service of the World Government, including Members of the World Parliament and of all other elective, appointive and civil service personnel, so that every person in the service of the World Government may have a better understanding of the functions, structure, procedures and inter-relationships of the various organs, departments, bureaus, offices, institutes, commissions, agencies and other parts of the World Government.

8.4.1.2 To prepare and conduct courses and seminars for information, education, discussion, updating and new ideas in all areas of world problems, particularly for Members of the World Parliament and of the World Executive, and for the chief personnel of all organs, departments and agencies of the World Government, but open to all in the service of the World Government.

8.4.1.3 To bring in qualified persons from private and public universities, colleges and research and action organizations of many countries, as well as other qualified persons, to lecture and to be resource persons for the courses and seminars organized by the Institute on Governmental Procedures and World Problems.

8.4.1.4 To contract with private or public universities and colleges or other agencies to conduct courses and seminars for the Institute.

8.4.2 The Institute on Governmental Procedures and World Problems shall be supervised by a ten member commission in addition to the Senior Administrator and Cabinet Minister or Vice President. The commission shall be composed of one commissioner each to be named by the House of Peoples, the House of Nations, the House of Counsellors, the Presidium, the Collegium of World Judges, The World Ombudsmus, The World Attorneys General Office, the Agency for Research and Planning, the Agency for Technological and Environmental Assessment, and the World Financial Administration. Commissioners shall serve five year terms, and may serve consecutive terms.

Section 8.5   The Agency for Research and Planning

8.5.1 The functions of the Agency for Research and Planning shall be as follows, but not limited thereto:

8.5.1.1 To serve the World Parliament, the World Executive, the World Administration, and other organs, departments and agencies of the World Government in any matter requiring research and planning within the competence of the agency.

8.5.1.2 To prepare and maintain a comprehensive inventory of world resources.

8.5.1.3 To prepare comprehensive long-range plans for the development, conservation, recycling and equitable sharing of the resources of Earth for the benefit of all people on Earth, subject to legislative action by the World Parliament.

8.5.1.4 To prepare and maintain a comprehensive list and description of all world problems, including their interrelationships, impact time projections and proposed solutions, together with bibliographies.

8.5.1.5 To do research and help prepare legislative measures at the request of any Member of the World Parliament or of any committee of any House of the World Parliament.

8.5.1.6 To do research and help prepare proposed legislation or proposed legislative programs and schedules at the request of the Presidium or Executive Cabinet or of any Cabinet Minister.

8.5.1.7 To do research and prepare reports at the request of any other organ, department or agency of the World Government.

8.5.1.8 To enlist the help of public and private universities, colleges, research agencies, and other associations and organizations for various research and planning projects.

8.5.1.9 To contract with public and private universities, colleges, research agencies and other organizations for the preparation of specific reports, studies and proposals.

8.5.1.10 To maintain a comprehensive World Library for the use of all Members of the World Parliament, and for the use of all other officials and persons in the service of the World Government, as well as for public information.

8.5.2 The Agency for Research and Planning shall be supervised by a ten member commission in addition to the Senior Administrator and Cabinet Minister or Vice President. The commission shall be composed of one commissioner each to be named by the House of Peoples, the House of Nations, the House of Counsellors, the Presidium, the Collegium of World Judges, the Office of World Attorneys General, World Ombudsmus, the Agency for Technological and Environmental Assessment, the Institute on Governmental Procedures and World Problems, and the World Financial Administration. Commissioners shall serve five year terms, and may serve consecutive terms.

Section 8.6   The Agency for Technological and Environmental Assessment

8.6.1 The functions of the agency for Technological and Environmental Assessment shall include the following, but not limited thereto:

8.6.1.1 To establish and maintain a registration and description of all significant technological innovations, together with impact projections.

8.6.1.2 To examine, analyze and assess the impacts and consequences of technological innovations which may have either significant beneficial or significant harmful or dangerous consequences for human life or for the ecology of life on
Earth, or which may require particular regulations or prohibitions to prevent or eliminate dangers or to assure benefits.

8.6.1.3 To examine, analyze and assess environmental and ecological problems, in particular the environmental and ecological
problems which may result from any intrusions or changes
of the environment or ecological relationships which may be
caused by technological innovations, processes of resource
development, patterns of human settlements, the production of energy, patterns of economic and industrial development, or other man-made intrusions and changes of the
environment, or which may result from natural causes.

8.6.1.4 To maintain a global monitoring network to measure possible harmful effects of technological innovations and environmental disturbances so that corrective measures can be
designed.

8.6.1.5 To prepare recommendations based on technological and environmental analyses and assessments, which can serve as
guides to the World Parliament, the World Executive, the
World Administration, the Agency for Research and Planning, and to the other organs, departments and agencies of
the World Government, as well as to individuals in the service of the World Government and to national and local governments and legislative bodies.

8.6.1.6 To enlist the voluntary or contractual aid and participation
of private and public universities, colleges, research institutions and other associations and organizations in the work
of technological and environmental assessment.

8.6.1.7 To enlist the voluntary or contractual aid and participation
of private and public universities and colleges, research institutions and other organizations in devising and developing alternatives to harmful or dangerous technologies and
environmentally disruptive activities, and in devising controls to assure beneficial results from technological innovations or to prevent harmful results from either technological

innovations or environmental changes, all subject to legislation for implementation by the World Parliament.

8.6.2 The Agency for Technological and Environmental Assessment shall be supervised by a ten member commission in addition to the Senior Administrator and Cabinet Minister or Vice President. The commission shall be composed of one commissioner from each of ten World Electoral and Administrative Magna-Regions. The persons to serve as commissioners shall be nominated by the House of Counsellors, and then appointed by the World Presidium for five year terms. Commissioners may serve consecutive terms.

Section 8.7   The World Financial Administration

8.7.1 The functions of the World Financial Administration shall include the following, but not limited thereto:

8.7.1.1 To establish and operate the procedures for the collection of revenues for the World Government, pursuant to legislation by the World Parliament, inclusive of taxes, globally accounted social and public costs, licenses, fees, revenue sharing arrangements, income derived from supra-national public enterprises or projects or resource developments, and all other sources.

8.7.1.2 To operate a Planetary Accounting Office, and thereunder to make cost/benefit studies and reports of the functioning and activities of the World Government and of its several organs, departments, branches, bureaus, offices, commissions, institutes, agencies and other parts or projects. In making such studies and reports, account shall be taken not only of direct financial costs and benefits, but also of human, social, environmental, indirect, long-term and other costs and benefits, and of actual or possible hazards and damages. Such studies and reports shall also be designed to uncover any wastes, inefficiencies, misapplications, corruptions, diversions, unnecessary costs, and other possible irregularities.

8.7.1.3 To make cost/benefit studies and reports at the request of any House or committee of the World Parliament, and of the Presidium, the Executive Cabinet, the World Ombudsmus, the Office of World Attorneys General, the World Supreme Court, or of any administrative department or any agency of the Integrative Complex, as well as upon its own initiative.

8.7.1.4 To operate a Planetary Comptrollers Office and thereunder to supervise the disbursement of the funds of the World Government for all purposes, projects and activities duly authorized by this World Constitution, the World Parliament, the World Executive, and other organs, departments and agencies of the World Government.

8.7.1.5 To establish and operate a Planetary Banking System, making the transition to a common global currency, under the terms of specific legislation passed by the World Parliament.

8.7.1.6 Pursuant to specific legislation enacted by the World Parliament, and in conjunction with the Planetary Banking System, to establish and implement the procedures of a Planetary Monetary and Credit System based upon useful productive capacity and performance, both in goods and services. Such a monetary and credit system shall be designed for use within the Planetary Banking System for the financing of the activities and projects of the World Government, and for all other financial purposes approved by the World Parliament, without requiring the payment of interest on bonds, investments or other claims of financial ownership or debt.

8.7.1.7 To establish criteria for the extension of financial credit based upon such considerations as people available to work, usefulness, cost/benefit accounting, human and social values, environmental health and esthetics, minimizing disparities, integrity, competent management, appropriate technology, potential production and performance.

8.7.1.8 To establish and operate a Planetary Insurance System in areas of world need which transcend national boundaries and in accordance with legislation passed by the World Parliament.

8.7.1.9 To assist the Presidium as may be requested in the technical preparation of budgets for the operation of the World Government.

8.7.2 The World Financial Administration shall be supervised by a commission of ten members, together with a Senior Administrator and a Cabinet Minister or Vice President. The commission shall be composed of one commissioner each to be named by the House of Peoples, the House of Nations, the House of Counsellors, the Presidium, the Collegium of World Judges, the Office of Attorneys General, the World Ombudsmus, the Agency for Research and Planning, the Agency for Technological and Environmental Assessment, and the Institute on Governmental Procedures and World Problems. Commissioners shall serve terms of five years, and may serve consecutive terms.

Section 8.8   Commission for Legistlative Review

8.8.1 The functions of the Commission for Legislative Review shall be to examine World Legislation and World Laws which the World Parliament enacts or adopts from the previous Body of International Law for the purpose of analyzing whether any particular legislation or law has become obsolete or obstructive or defective in serving the purposes intended; and to make recommendations to the World Parliament accordingly for repeal or amendment or replacement.

8.8.2 The Commission for Legislative Review shall be composed of twelve members, including two each to be elected by the House of Peoples, the House of Nations, the House of Counsellors, the Collegium of World Judges, the World Ombudsmus and the Presidium. Members of the Commission shall serve terms of ten years, and may be re-elected to serve consecutive terms. One half of the Commission members after the Commission is first formed shall be elected every five years, with the first terms for one half of the members to be only five years.

Article 9   The World Judiciary

Section 9.1   Jurisdiction of the World Supreme Court

9.1.1 A World Supreme Court shall be established, together with such regional and district World Courts as may subsequently be found necessary. The World Supreme Court shall comprise a number of benches.

9.1.2 The World Supreme Court, together with such regional and district World Courts as may be established, shall have mandatory jurisdiction in all cases, actions, disputes, conflicts, violations of law, civil suits, guarantees of civil and human rights, constitutional interpretations, and other litigations arising under the provisions of this World Constitution, world legislation, and the body of world law approved by the World Parliament.

9.1.3 Decisions of the World Supreme Court shall be binding on all parties involved in all cases, actions and litigations brought before any bench of the World Supreme Court for settlement. Each bench of the World Supreme Court shall constitute a court of highest appeal, except when matters of extra-ordinary public importance are assigned or transferred to the Superior Tribunal of the World Supreme Court, as provided in Section 5 of Article 9.

Section 9.2   Benches of the World Supreme Court

The benches of the World Supreme Court and their respective jurisdictions shall be as follows:

9.2.1 Bench for Human Rights: To deal with issues of human rights arising under the guarantee of civil and human rights provided by Article 12 of this World Constitution, and arising in pursuance of the provisions of Article 13 of this World Constitution, and arising otherwise under world legislation and the body of world law approved by the World Parliament.

9.2.2 Bench for Criminal Cases: To deal with issues arising from the violation of world laws and world legislation by individuals, corporations, groups and associations, but not issues primarily concerned with human rights.

9.2.3 Bench for Civil Cases: To deal with issues involving civil law suits and disputes between individuals, corporations, groups and associations arising under world legislation and world law and the administration thereof.

9.2.4 Bench for Constitutional Cases: To deal with the interpretation of the World Constitution and with issues and actions arising in connection with the interpretation of the World Constitution.

9.2.5 Bench for International Conflicts: To deal with disputes, conflicts and legal contest arising between or among the nations which have joined in the Federation of Earth.

9.2.6 Bench for Public Cases: To deal with issues not under the jurisdiction of another bench arising from conflicts, disputes, civil suits or other legal contests between the World Government and corporations, groups or individuals, or between national governments and corporations, groups or individuals in cases involving world legislation and world law.

9.2.7 Appellate Bench: To deal with issues involving world legislation and world law which may be appealed from national courts; and to decide which bench to assign a case or action or litigation when a question or disagreement arises over the proper jurisdiction.

9.2.8 Advisory Bench: To give opinions upon request on any legal question arising under world law or world legislation, exclusive of contests or actions involving interpretation of the World Constitution. Advisory opinions may be requested by any House or committee of the World Parliament, by the Presidium, any Administrative Department, the Office of World Attorneys General, the World Ombudsmus, or by any agency of the Integrative Complex.

9.2.9 Other benches may be established, combined or terminated upon recommendation of the Collegium of World Judges with approval by the World Parliament; but benches number one through eight may not be combined nor terminated except by amendment of this World Constitution.

Section 9.3   Seats of the World Supreme Court

9.3.1 The primary seat of the World Supreme Court and all benches shall be the same as for the location of the Primary World Capital and for the location of the World Parliament and the World Executive.

9.3.2 Continental seats of the World Supreme Court shall be established in the four secondary capitals of the World Government located in four different Continental Divisions of Earth, as provided in Article 15.

9.3.3 The following permanent benches of the World Supreme Court shall be established both at the primary seat and at each of the continental seats: Human Rights, Criminal Cases, Civil Cases, and Public Cases.

9.3.4 The following permanent benches of the World Supreme Court shall be located only at the primary seat of the World Supreme Court: Constitutional Cases, International Conflicts, Appellate Bench, and Advisory Bench.

9.3.5 Benches which are located permanently only at the primary seat of the World Supreme Court may hold special sessions at the other continental seats of the World Supreme Court when necessary, or may establish continental circuits if needed.

9.3.6 Benches of the World Supreme Court which have permanent continental locations may hold special sessions at other locations when needed, or may establish regional circuits if needed.

Section 9.4   The Collegium of World Judges

9.4.1 A Collegium of World Judges shall be established by the World Parliament. The Collegium shall consist of a minimum of twenty member judges, and may be expanded as needed but not to exceed sixty members.

9.4.2 The World Judges to compose the Collegium of World Judges shall be nominated by the House of Counsellors and shall be elected by plurality vote of the three Houses of the World Parliament in joint session. The House of Counsellors shall nominate between two and three times the number of world judges to be elected at any one time. An equal number of World Judges shall be elected from each of ten World Electoral and Administrative Magna-Regions, if not immediately then by rotation.

9.4.3 The term of office for a World Judge shall be ten years. Successive terms may be served without limit.

9.4.4 The Collegium of World Judges shall elect a Presiding Council of World Judges, consisting of a Chief Justice and four Associate Chief Justices. One member of the Presiding Council of World Judges shall be elected from each of five Continental Divisions of Earth. Members of the Presiding Council of World Judges shall serve five year terms on the Presiding Council, and may serve two successive terms, but not two successive terms as Chief Justice.

9.4.5 The Presiding Council of World Judges shall assign all World Judges, including themselves, to the several benches of the World Supreme Court. Each bench for a sitting at each location shall have a minimum of three World Judges, except that the number of World Judges for benches on Continental Cases and International Conflicts, and the Appellate Bench, shall be no less than five.

9.4.6 The member judges of each bench at each location shall choose annually a Presiding Judge, who may serve two successive terms.

9.4.7 The members of the several benches may be reconstituted from time to time as may seem desirable or necessary upon the decision of the Presiding Council of World Judges. Any decision to re-constitute a bench shall be referred to a vote of the entire Collegium of World Judges by request of any World Judge.

9.4.8 Any World Judge may be removed from office for cause by an absolute two thirds majority vote of the three Houses of the World Parliament in joint session.

9.4.9 Qualifications for Judges of the World Supreme Court shall be at least ten years of legal or juristic experience, minimum age of thirty years, and evident competence in world law and the humanities.

9.4.10 The salaries, expenses, remunerations and prerogatives of the World Judges shall be determined by the World Parliament, and shall be reviewed every five years, but shall not be changed to the disadvantage of any World Judge during a term of office. All members of the Collegium of World Judges shall receive the same salaries, except that additional compensation may be given to the Presiding Council of World Judges.

9.4.11 Upon recommendation by the Collegium of World Judges, the World Parliament shall have the authority to establish regional and district world courts below the World Supreme Court, and to establish the jurisdictions thereof, and the procedures for appeal to the World Supreme Court or to the several benches thereof.

9.4.12 The detailed rules of procedure for the functioning of the World Supreme Court, the Collegium of World Judges, and for each bench of the World Supreme Court, shall be decided and amended by absolute majority vote of the Collegium of World Judges.

Section 9.5   The Superior Tribunal of World Supreme Court

9.5.1 A Superior Tribunal of the World Supreme Court shall be established to take cases which are considered to be of extra-ordinary public importance. The Superior Tribunal for any calendar year shall consist of the Presiding Council of World Judges together with one World Judge named by the Presiding Judge of each bench of the World Court sitting at the primary seat of the World Supreme Court. The composition of the Superior Tribunal may be continued unchanged for a second year by decision of the Presiding Council of World Judges.

9.5.2 Any party to any dispute, issue, case or litigation coming under the jurisdiction of the World Supreme Court, may apply to any particular bench of the World Supreme Court or to the Presiding Council of World Judges for the assignment or transfer of the case to the Superior Tribunal on the grounds of extra-ordinary public importance. If the application is granted, the case shall be heard and disposed of by the Superior Tribunal. Also, any bench taking any particular case, if satisfied that the case is of extra-ordinary public importance, may of its own discretion transfer the case to the Superior Tribunal.

Article 10   The Enforcement System

Section 10.1   Basic Principles

10.1.1 The enforcement of world law and world legislation shall apply directly to individual, and individuals shall be held responsible for compliance with world law and world legislation regardless of whether the individuals are acting in their own capacity or as agents or officials of governments at any level or of the institutions of governments, or as agents or officials of corporations, organizations, associations or groups of any kind.

10.1.2 When world law or world legislation or decisions of the world courts are violated, the Enforcement System shall operate to identify and apprehend the individuals responsible for violations.

10.1.3 Any enforcement action shall not violate the civil and human rights guaranteed under this World Constitution.

10.1.4 The enforcement of world law and world legislation shall be carried out in the context of a non-military world federation wherein all member nations shall disarm as a condition for joining and benefiting from the world federation, subject to Article 17, Sec. 3.8 and 4.6 The Federation of Earth and World Government under this World Constitution shall neither keep nor use weapons of mass destruction.

10.1.5 Those agents of the enforcement system whose function shall be to apprehend and bring to court violators of world law and world legislation shall be equipped only with such weapons as are appropriate for the apprehension of the individuals responsible for violations.

10.1.6 The enforcement of world law and world legislation under this World Constitution shall be conceived and developed primarily as the processes of effective design and administration of world law and world legislation to serve the welfare of all people on Earth, with equity and justice for all, in which the resources of Earth and the funds and the credits of the World Government are used only to serve peaceful human needs, and none used for weapons of mass destruction or for war making capabilities.

Section 10.2   The Structure for Enforcement: World Attorneys General

10.2.1 The Enforcement System shall be headed by an Office of World Attorneys General and a Commission of Regional World Attorneys.

10.2.2 The Office of World Attorneys General shall be comprised of five members, one of whom shall be designated as the World Attorney General and the other four shall each be designated an Associate World Attorney General.

10.2.3 The Commission of Regional World Attorneys shall consist of twenty Regional World Attorneys.

10.2.4 The members to compse the Office of World Attorneys General shall be nominated by the House of Counsellors, with three nominees from each Continental Division of Earth. One member of the Office shall be elected from each of five Continental Divisions by plurality vote of the three houses of the World Parliament in joint session.

10.2.5 The term of office for a member of the Office of World Attorneys General shall be ten years. A member may serve two consecutive terms. The position of World Attorney General shall rotate every two years among the five members of the Office. The order of rotation shall be decided among the five members of the Office.

10.2.6 The Office of World Attorneys General shall nominate members for the Commission of twenty Regional World Attorneys from the twenty World Electoral and Administrative Regions, with between two and three nominees submitted for each Region. From these nominations, the three Houses of the World Parliament in joint session shall elect one Regional World Attorney from each of the twenty Regions. Regional World Attorneys shall serve terms of five years, and may serve three consecutive terms.

10.2.7 Each Regional World Attorney shall organize and be in charge of an Office of Regional World Attorney. Each Associate World Attorney General shall supervise five Offices of Regional World Attorneys.

10.2.8 The staff to carry out the work of enforcement, in addition to the five members of the Office of World Attorneys General and the twenty Regional World Attorneys, shall be selected from civil service lists, and shall be organized for the following functions:

10.2.8.1 Investigation.

10.2.8.2 Apprehension and arrest.

10.2.8.3 Prosecution.

10.2.8.4 Remedies and correction.

10.2.8.5 Conflict resolution.

10.2.9 Qualifications for a member of the Office of World Attorneys General and for the Regional World Attorneys shall be at least thirty years of age, at least seven years legal experience, and education in law and the humanities.

10.2.10 The World Attorney General, the Associate World Attorneys General, and the Regional World Attorneys shall at all times be responsible to the World Parliament. Any member of the Office of World Attorneys General and any Regional World Attorney can be removed from office for cause by a simple majority vote of the three Houses of the World Parliament in joint session.

Section 10.3   The World Police

10.3.1 That section of the staff of the Office of World Attorneys General and of the Offices of Regional World Attorneys responsible for the apprehension and arrest of violators of world law and world legislation, shall be designated as World Police.

10.3.2 Each regional staff of the World Police shall be headed by a Regional World Police Captain, who shall be appointed by the Regional World Attorney.

10.3.3 The Office of World Attorneys General shall appoint a World Police Supervisor, to be in charge of those activities which transcend regional boundaries. The World Police Supervisor shall direct the Regional World Police Captains in any actions which require coordinated or joint action transcending regional boundaries, and shall direct any action which requires initiation or direction from the Office of World Attorneys General.

10.3.4 Searches and arrests to be made by World Police shall be made only upon warrants issued by the Office of World Attorneys General or by a Regional World Attorney.

10.3.5 World Police shall be armed only with weapons appropriate for the apprehension of the individuals responsible for violation of world law.

10.3.6 Employment in the capacity of World Police Captain and World Police Supervisor shall be limited to ten years.

10.3.7 The World Police Supervisor and any Regional World Police Captain may be removed from office for cause by decision of the

Office of World Attorneys General or by absolute majority vote of the three Houses of the World Parliament in joint session.

Section 10.4   The Means of Enforcement

10.4.1 Non-military means of enforcement of world law and world legislation shall be developed by the World Parliament and by the Office of World Attorneys General in consultation with the Commission of Regional World Attorneys, the Collegium of World Judges, the World Presidium, and the World Ombudsmus. The actual means of enforcement shall require legislation by the World Parliament.

10.4.2 Non-military means of enforcement which can be developed may include: Denial of financial credit; denial of material resources and personnel; revocation of licenses, charters, or corporate rights; impounding of equipment; fines and damage payments; performance of work to rectify damages; imprisonment or isolation; and other means appropriate to the specific situations.

10.4.3 To cope with situations of potential or actual riots, insurrection and resort to armed violence, particular strategies and methods shall be developed by the World Parliament and by the Office of World Attorneys General in consultation with the Commission of Regional World Attorneys, the collegium of World Judges, the Presidium and the World Ombudsmus. Such strategies and methods shall require enabling legislation by the World Parliament where required in addition to the specific provisions of this World Constitution.

10.4.4 A basic condition for preventing outbreaks of violence which the Enforcement System shall facilitate in every way possible, shall be to assure a fair hearing under non-violent circumstances for any person or group having a grievance, and likewise to assure a fair opportunity for a just settlement of any grievance with due regard for the rights and welfare of all concerned.

Article 11   The World Ombudsmus

Section 11.1   Functions and Powers of the World Ombudsmus

The functions and powers of the World Ombudsmus, as public defender, shall include the following:

11.1.1 To protect the People of Earth and all individuals against violations or neglect of universal human and civil rights which are stipulated in Article 12 and other sections of this World Constitution.

11.1.2 To protect the People of Earth against violations of this World Constitution by any official or agency of the World Government, including both elected and appointed officials or public employees regardless of organ, department, office, agency or rank.

11.1.3 To press for the implementation of the Directive Principles for the World Government as defined in Article 13 of this World Constitution.

11.1.4 To promote the welfare of the people of Earth by seeking to assure that conditions of social justice and of minimizing disparities are achieved in the implementation and administration of world legislation and world law.

11.1.5 To keep on the alert for perils to humanity arising from technological innovations, environmental disruptions and other diverse sources, and to launch initiatives for correction or prevention of such perils.

11.1.6 To ascertain that the administration of otherwise proper laws, ordinances and procedures of the World Government do not result in unforseen injustices or inequities, or become stultified in bureaucracy or the details of administration.

11.1.7 To receive and hear complaints, grievances or requests for aid from any person, group, organization, association, body politic or agency concerning any matter which comes within the purview of the World Ombudsmus.

11.1.8 To request the Office of World Attorneys General or any Regional World Attorney to initiate legal actions or court proceedings whenever and wherever considered necessary or desirable in the view of the World Ombudsmus.

11.1.9 To directly initiate legal actions and court proceedings whenever the World Ombudsmus deems necessary.

11.1.10 To review the functioning of the departments, bureaus, offices, commissions, institutes, organs and agencies of the World Government to ascertain whether the procedures of the World government are adequately fulfilling their purposes and serving the welfare of humanity in optimum fashion, and to make recommendations for improvements.

11.1.11 To present an annual report to the World Parliament and to the Presidium on the activities of the World Ombudsmus, together with any recommendations for legislative measures to improve the functioning of the World Government for the purpose of better serving the welfare of the People of Earth.

Section 11.2   Composition of the World Ombudsmus

11.2.1 The World Ombudsmus shall be headed by a Council of World Ombudsen of five members, one of whom shall be designated as Principal World Ombudsan, while the other four shall each be designated as an Associate World Ombudsan.

11.2.2 Members to compose the Council of World Ombudsen shall be nominated by the House of Counsellors, with three nominees from each Continental Division of Earth. One member of the Council shall be elected from each of five Continental Divisions by plurality vote of the three Houses of the World Parliament in joint session.

11.2.3 The term of office for a World Ombudsan shall be ten years. A World Ombudsan may serve two successive terms. The position of Principal World Ombudsan shall be rotated every two years. The order of rotation shall be determined by the Council of World Ombudsen.

11.2.4 The Council of World Ombudsen shall be assisted by a Commission of World Advocates of twenty members. Members for the Commission of World Advocates shall be nominated by the Council of World Ombudsen from twenty World Electoral and Administrative Regions, with between two and three nominees submitted for each Region. One World Advocate shall be elected from each of the twenty World Electoral and Administrative Regions by the three Houses of the World Parliament in joint session. World Advocates shall serve terms of five years, and may serve a maximum of four successive terms.

11.2.5 The Council of World Ombudsen shall establish twenty regional offices, in addition to the principal world office at the primary seat of the World Government. The twenty regional offices of the World Ombudsmus shall parallel the organization of the twenty Offices of Regional World Attorney.

11.2.6 Each regional office of the World Ombudsmus shall be headed by a World Advocate. Each five regional offices of the World Ombudmus shall be supervised by an Associate World Ombudsan.

11.2.7 Any World Ombudsan and any World Advocate may be removed from office for cause by an absolute majority vote of the three Houses of the World Parliament in joint session.

11.2.8 Staff members for the World Ombudsmus and for each regional office of the World Ombudsmus shall be selected and employed from civil service lists.

11.2.9 Qualifications for World Ombudsan and for World Advocate shall be at least thirty years of age, at least five years legal experience, and education in law and other relevant education.

Article 12   Bill of Rights for the Citizens of Earth

12.1 Equal rights for all citizens of the Federation of Earth, with no discrimination on grounds of race, color, caste, nationality, sex, religion, political affiliation, property, or social status.

12.2 Equal protection and application of world legislation and world laws for all citizens of the Federation of Earth.

12.3 Freedom of thought and conscience, speech, press, writing, communication, expression, publication, broadcasting, telecasting, and cinema, except as an overt part of or incitement to violence, armed riot or insurrection.

12.4 Freedom of assembly, association, organization, petition and peaceful demonstration.

12.5 Freedom to vote without duress, and freedom for political organization and campaigning without censorship or recrimination.

12.6 Freedom to profess, practice and promote religion or religious beliefs or no religion or religious belief.

12.7 Freedom to profess and promote political beliefs or no political beliefs.

12.8 Freedom for investigation, research and reporting.

12.9 Freedom to travel without passport or visas or other forms of registration used to limit travel between, among or within nations.

12.10 Prohibition against slavery, peonage, involuntary servitude, and conscription of labor.

12.11 Prohibition against military conscription.

12.12 Safety of person from arbitrary or unreasonable arrest, detention, exile, search or seizure; requirement of warrants for searches and arrests.

12.13 Prohibition against physical or psychological duress or torture during any period of investigation, arrest, detention or imprisonment, and against cruel or unusual punishment.

12.14 Right of habeous corpus; no ex-post-facto laws; no double jeopardy; right to refuse self-incrimination or the incrimination of another.

12.15 Prohibition against private armies and paramilitary organizations as being threats to the common peace and safety.

12.16 Safety of property from arbitrary seizure; protection against exercise of the power of eminent domain without reasonable compensation.

12.17 Right to family planning and free public assistance to achieve family planning objectives.

12.18 Right of privacy of person, family and association; prohibition against surveillance as a means of political control.

Article 13   Directive Principles for the Earth Federation

It shall be the aim of the World Government to secure certain other rights for all inhabitants within the Federation of Earth, but without immediate guarantee of universal achievement and enforcement. These rights are defined as Directive Principles, obligating the World Government to pursue every reasonable means for universal realization and implementation, and shall include the following:

13.1 Equal opportunity for useful employment for everyone, with wages or remuneration sufficient to assure human dignity.

13.2 Freedom of choice in work, occupation, employment or profession.

13.3 Full access to information and to the accumulated knowledge of the human race.

13.4 Free and adequate public education available to everyone, extending to the pre-university level; Equal opportunities for elementary and higher education for all persons; equal opportunity for continued education for all persons throughout life; the right of any person or parent to choose a private educational institution at any time.

13.5 Free and adequate public health services and medical care available to everyone throughout life under conditions of free choice.

13.6 Equal opportunity for leisure time for everyone; better distribution of the work load of society so that every person may have equitable leisure time opportunities.

13.7 Equal opportunity for everyone to enjoy the benefits of scientific and technological discoveries and developments.

13.8 Protection for everyone against the hazards and perils of technological innovations and developments.

13.9 Protection of the natural environment which is the common heritage of humanity against pollution, ecological disruption or damage which could imperil life or lower the quality of life.

13.10 Conservation of those natural resources of Earth which are limited so that present and future generations may continue to enjoy life on the planet Earth.

13.11 Assurance for everyone of adequate housing, of adequate and nutritious food supplies, of safe and adequate water supplies, of pure air with protection of oxygen supplies and the ozone layer, and in general for the continuance of an environment which can sustain healthy living for all.

13.12 Assure to each child the right to the full realization of his or her potential.

13.13 Social Security for everyone to relieve the hazards of unemployment, sickness, old age, family circumstances, disability, catastrophies of nature, and technological change, and to allow retirement with sufficient lifetime income for living under conditions of human dignity during older age.

13.14 Rapid elimination of and prohibitions against technological hazards and manmade environmental disturbances which are found to create dangers to life on Earth.

13.15 Implementation of intensive programs to discover, develop and institute safe alternatives and practical substitutions for technologies which must be eliminated and prohibited because of hazards and dangers to life.

13.16 Encouragement for cultural diversity; encouragement for decentralized administration.

13.17 Freedom for peaceful self-determination for minorities, refugees and dissenters.

13.18 Freedom for change of residence to anywhere on Earth conditioned by provisions for temporary sanctuaries in events of large numbers of refugees, stateless persons, or mass migrations.

13.19 Prohibition against the death penalty.

Article 14 Safeguards and Reservations

Section 14.1 Certain Safeguards

The World Government shall operate to secure for all nations and peoples within the Federation of Earth the safeguards which are defined hereunder:

14.1.1 Guarantee that full faith and credit shall be given to the public acts, records, legislation and judicial proceedings of the member nations within the Federation of Earth, consistent with the several provisions of this World Constitution.

14.1.2 Assure freedom of choice within the member nations and countries of the Federation of Earth to determine their internal political, economic and social systems consistent with the guarantees and protections given under this World Constitution to assure civil liberties and human rights and a safe environment for life, and otherwise consistent with the several provisions of this World Constitution.

14.1.3 Grant the right of asylum within the Federation of Earth for persons who may seek refuge from countries or nations which are not yet included within the Federation of Earth.

14.1.4 Grant the right of individuals and groups, after the Federation of Earth includes 90 percent of the territory of Earth, to peacefully leave the hegemony of the Federation of Earth and to live in suitable territory set aside by the Federation neither restricted nor protected by the World Government, provided that such territory does not extend beyond five percent of Earth's habitable territory, is kept completely disarmed and not used as a base for inciting violence or insurrection within or against the Federation of Earth or any member nation, and is kept free of acts of environmental or technological damage which seriously affect Earth outside such territory.

Section 14.2 Reservation of Powers

The powers not delegated to the World Government by this World Constitution shall be reserved to the nations of the Federation of Earth and to the people of Earth.

Article 15 World Federal Zones and the World Capitals

Section 15.1 Word Federal Zones

15.1.1 Twenty World Federal Zones shall be established within the twenty World Electoral and Administrative Regions, for the purposes of the location of the several organs of the World Government and of the administrative departments, the world courts, the offices of the Regional World Attorneys, the offices of the World Advocates, and for the location of other branches, departments, institutes, offices, bureaus, commissions, agencies and parts of the World Government.

15.1.2 The World Federal Zones shall be established as the needs and resources of the World Government develop and expand. World Federal Zones shall be established first within each of five Continental Divisions.

15.1.3 The location and administration of the World Federal Zones, including the first five, shall be determined by the World Parliament.

Section 15.2 The World Capitals

15.2.1 Five World Capitals shall be established in each of five Continental Divisions of Earth, to be located in each of the five World Federal Zones which are established first as provided in Article 15 of this World Constitution.

15.2.2 One of the World Capitals shall be designated by the World Parliament as the Primary World Capital, and the other four shall be designated as Secondary World Capitals.

15.2.3 The primary seats of all organs of the World Government shall be located in the Primary World Capital, and other major seats of the several organs of the World Government shall be located in the Secondary World Capitals.

Section 15.3   Locational Procedures

15.3.1 Choices for location of the twenty World Federal Zones and for the five World Capitals shall be proposed by the Presidium, and then shall be decided by a simple majority vote of the three Houses of the World Parliament in joint session. The Presidium shall offer choices of two or three locations in each of the twenty World Electoral and Administrative Regions to be World Federal Zones, and shall offer two alternative choices for each of the five World Capitals.

15.3.2 The Presidium in consultation with the Executive Cabinet shall then propose which of the five World Capitals shall be the Primary World Capital, to be decided by a simply majority vote of the three Houses of the World Parliament in joint session.

15.3.3 Each organ of the World Government shall decide how best to apportion and organize its functions and activities among the five World Capitals, and among the twenty World Federal Zones, subject to specific directions from the World Parliament.

15.3.4 The World Parliament may decide to rotate its sessions among the five World Capitals, and if so, to decide the procedure for rotation.

15.3.5 For the first two operative stages of World Government as defined in Article 17, and for the Provisional World Government as defined in Article 19, a provisional location may be selected for the Primary World Capital. The provisional location need not be continued as a permanent location.

15.3.6 Any World Capital or World Federal Zone may be relocated by an absolute two-thirds majority vote of the three Houses of the World Parliament in joint session.

15.3.7 Additional World Federal Zones may be designated if found necessary by proposal of the Presidium and approval by an absolute majority vote of the three Houses of the World Parliament in joint session.

Article 16   World Territories and Exterior Relations

Section 16.1   World Territory

16.1.1 Those areas of the Earth and Earth's moon which are not under the jurisdiction of existing nations at the time of forming the Federation of Earth, or which are not reasonably within the province of national ownership and administration, or which are declared to be World Territory subsequent to establishment of the Federation of Earth, shall be designated as World Territory and shall belong to all of the people of Earth.

16.1.2 The administration of World Territory shall be determined by the World Parliament and implemented by the World Executive, and shall apply to the following areas:

16.1.2.1 All oceans and seas having an international or supranational character, together with the seabeds and resources thereof, beginning at a distance of twenty kilometers offshore, excluding inland seas of traditional national ownership.

16.1.2.2 Vital straits, channels, and canals.

16.1.2.3 The atmosphere enveloping Earth, beginning at an elevation of one kilometer above the general surface of the land, excluding the depressions in areas of much variation in elevation.

16.1.2.4 Man-made satellites and Earth's moon.

16.1.2.5 Colonies which may choose the status of World Territory; non-independent territories under the trust administration of nations or of the United Nations; any islands or atolls which are unclaimed by any nation; independent lands or countries which choose the status of World Territory; and disputed lands which choose the status of World Territory.

16.1.3 The residents of any World Territory, except designated World Federal Zones, shall have the right within reason to decide by plebiscite to become a self-governing nation within the Federation of Earth, either singly or in combination with other World Territories, or to unite with an existing nation with the Federation of Earth.

Section 16.2 Exterior Relations

16.2.1 The World Government shall maintain exterior relations with those nations of Earth which have not joined the Federation of Earth. Exterior relations shall be under the administration of the Presidium, subject at all times to specific instructions and approval by the World Parliament.

16.2.2 All treaties and agreements with nations remaining outside the Federation of Earth shall be negotiated by the Presidium and must be ratified by a simple majority vote of the three Houses of the World Parliament.

16.2.3 The World Government for the Federation of Earth shall establish and maintain peaceful relations with other planets and celestial bodies where and when it may become possible to establish communications with the possible inhabitants thereof.

16.2.4 All explorations into outer space, both within and beyond the solar system in which Planet Earth is located, shall be under the exclusive direction and control of the World Government, and shall be conducted in such manner as shall be determined by the World Parliament.

Article 17 Ratification and Implementation

Section 17.1 Ratification of the World Constitution

17.1.1 The World Constitution shall be transmitted to the General Assembly of the United Nations Organization and to each national government on Earth, with the request that the World Constitution be submitted to the national legislature of each nation for preliminary ratification and to the people of each nation for final ratification by popular referendum.

17.1.2 Preliminary ratification by a national legislature shall be accomplished by simple majority vote of the national legislature.

17.1.3 Final ratification by the people shall be accomplished by a simple majority of votes cast in a popular referendum, provided that a minimum of twenty-five percent of eligible voters of age eighteen years and over have cast ballots within the nation or country or within World Electoral and Administrative Districts.

17.1.4 In the case of a nation without a national legislature, the head of the national government shall be requested to give preliminary ratification and to submit the World Constitution for final ratification by popular referendum.

17.1.5 In the event that a national government, after six months, fails to submit the World Constitution for ratification as requested, then the global agency assuming responsibility for the worldwide ratification campaign may proceed to conduct a direct referendum for ratification of the World Constitution by the people. Direct referendums may be organized on the basis of entire nations or countries, or on the basis of existing defined communities within nations.

17.1.6 In the event of a direct ratification referendum, final ratification shall be accomplished by a majority of the votes cast whether for an entire nation or for a World Electoral and Administrative District, provided that ballots are cast by a minimum of twenty-five percent of eligible voters of the area who are over eighteen years of age.

17.1.7 For ratification by existing communities within a nation, the procedure shall be to request local communities, cities, counties, states, provinces, cantons, prefectures, tribal jurisdictions, or other defined political units within a nation to ratify the World Constitution, and to submit the World Constitution for a referendum vote by the citizens of the community or political unit. Ratification may be accomplished by proceeding in this way until all eligible voters of age eighteen and above within the nation or World Electoral and Administrative District have had the opportunity to vote, provided that ballots are cast by a minimum of twenty-five percent of those eligible to vote.

17.1.8 Prior to the Full Operative Stage of World Government, as defined under Section 5 of Article 17, the universities, colleges and scientific academies and institutes in any country may ratify the World Constitution, thus qualifying them for participation in the nomination of Members of the World Parliament to the House of Counsellors.

17.1.9 In the case of those nations currently involved in serious international disputes or where traditional enmities and chronic disputes may exist among two or more nations, a procedure for concurrent paired ratification shall be instituted whereby the nations which are parties to a current or chronic international dispute or conflict may simultaneously ratify the World Constitution. In such cases, the paired nations shall be admitted into the Federation of Earth simultaneously, with the obligation for each such nation to immediately turn over all weapons of mass destruction to the World Government, and to turn over the conflict or dispute for mandatory peaceful settlement by the World Government.

17.1.10 Each nation or political unit which ratifies this World Constitution, either by preliminary ratification or final ratification, shall be bound never to use any armed forces or weapons of mass destruction against another member or unit of the Federation of Earth, regardless of how long it may take to achieve full disarmament of all the nations and political units which ratify this World Constitution.

17.1.11 When ratified, the Constitution for the Federation of Earth becomes the supreme law of Earth. By the act of ratifying this Earth Constitution, any provision in the Constitution or Legislation of any country so ratifying, which is contrary to this Earth Constitution, is either repealed or amended to conform with the Constitution for the Federation of Earth, effective as soon as 25 countries have so ratified. The amendment of National or State Constitutions to allow entry into World Federation is not necessary prior to ratification of the Constitution for the Federation of Earth.

Section 17.2   Stages of Implementation

17.2.1 Implementation of this World Constitution and the establishment of World Government pursuant to the terms of this World Constitution, may be accomplished in three stages, as follows, in addition to the stage of a Provisional World Government as provided under Article 19:

17.2.1.1 First Operative Stage of World Government.

17.2.1.2 Second Operative Stage of World Government.

17.2.1.3 Full Operative Stage of World Government.

17.2.2 At the beginning and during each stage, the World Parliament and the World Executive together shall establish goals and develop means for the progressive implementation of the World

Constitution, and for the implementation of legislation enacted by the World Parliament.

Section 17.3   First Operative Stage of World Government

17.3.1 The first operative stage of World Government under this World Constitution shall be implemented when the World Constitution is ratified by a sufficient number of nations and/or people to meet one or the other of the following conditions or equivalent:

17.3.1.1 Preliminary or final ratification by a minimum of twenty-five nations, each having a population of more than 100,000.

17.3.1.2 Preliminary or final ratification by a minimum of ten nations above 100,000 population, together with ratification by direct referendum within a minimum of fifty additional World Electoral and Administrative Districts.

17.3.1.3 Ratification by direct referendum within a minimum of 100 World Electoral and Administrative Districts, even though no nation as such has ratified.

17.3.2 The election of Members of the World Parliament to the House of Peoples shall be conducted in all World Electoral and Administrative Districts where ratification has been accomplished by popular referendum.

17.3.3 The election of Members of the World Parliament to the House of Peoples may proceed concurrently with direct popular referendums both prior to and after the First Operative Stage of World Government is reached.

17.3.4 The appointment or election of Members of the World Parliament to the House of Nations shall proceed in all nations where preliminary ratification has been accomplished.

17.3.5 One-fourth of the Members of the World Parliament to the House of Counsellors may be elected from nominees submitted by universities and colleges which have ratified the World Constitution.

17.3.6 The World Presidium and the Executive Cabinet shall be elected according to the provisions in Article 6, except that in the absence of a House of Counsellors, the nominations shall be made by the members of the House of Peoples and of the House of Nations in joint session. Until this is accomplished, the Presidium and Executive Cabinet of the Provisional World Government as provided in Article 19, shall continue to serve.

17.3.7 When composed, the Presidium for the first operative stage of World Government shall assign or re-assign Ministerial posts among Cabinet and Presidium members, and shall immediately establish or confirm a World Disarmament Agency and a World Economic and Development Organization.

17.3.8 Those nations which ratify this World Constitution and thereby join the Federation of Earth, shall immediately transfer all weapons of mass destruction as defined and designated by the World Disarmament Agency to that Agency. (See Article 19, Sections A-2-d, B-6 and E-5). The World Disarmament Agency shall immediately immobilize all such weapons and shall proceed with dispatch to dismantle, convert to peacetime use, re-cycle the materials thereof or otherwise destroy all such weapons. During the first operative stage of World Government, the ratifying nations may retain armed forces equipped with weapons other than weapons of mass destruction as defined and designated by the World Disarmament Agency.

17.3.9 Concurrently with the reduction or elimination of such weapons of mass destruction and other military expenditures as can be accomplished during the first operative stage of World Government, the member nations of the Federation of Earth shall pay annually to the Treasury of the World Government amounts equal to one-half the amounts saved from their respective national military budgets during the last year before joining the Federation, and shall continue such payments until the full operative stage of World Government is reached. The World Government shall use fifty percent of the funds thus received to finance the work and projects of the World Economic Development Organization.

17.3.10 The World Parliament and the World Executive shall continue to develop the organs, departments, agencies and activities originated under the Provisional World Government, with such amendments as deemed necessary; and shall proceed to establish and beg in the following organs, departments and agencies of the

World Government, if not already underway, together with such other departments, and agencies as are considered desirable and feasible during the first operative stage of World Government:

17.3.10.1 The World Supreme Court;

17.3.10.2 The Enforcement System;

17.3.10.3 The World Ombudsmus;

17.3.10.4 The World Civil Service Administration;

17.3.10.5 The World Financial Administration;

17.3.10.6 The Agency for Research and Planning;

17.3.10.7 The Agency for Technological and Environmental Assessment;

17.3.10.8 An Emergency Earth Rescue Administration, concerned with all aspects of climate change and related factors;

17.3.10.9 An Integrated Global Energy System, based on environmentally safe sources;

17.3.10.10 A World University System, under the Department of Education;

17.3.10.11 A World Corporations Office, under the Department of Commerce and Industry;

17.3.10.12 The World Service Corps;

17.3.10.13 A World Oceans and Seabeds Administration.

17.3.11 At the beginning of the first operative stage, the Presidium in consultation with the Executive Cabinet shall formulate and put forward a proposed program for solving the most urgent world problems currently confronting humanity.

17.3.12 The World Parliament shall proceed to work upon solutions to world problems. The World Parliament and the World Executive working together shall institute through the several organs, departments and agencies of the World Government whatever means shall seem appropriate and feasible to accomplish the implementation and enforcement of world legislation, world law and the World Constitution; and in particular shall take certain decisive actions for the welfare of all people on Earth, applicable throughout the world, including but not limited to the following:

17.3.12.1 Expedite the organization and work of an Emergency Earth Rescue Administration, concerned with all aspects of climate change and climate crises;

17.3.12.2 Expedite the new finance, credit and monetary system, to serve human needs;

17.3.12.3 Expedite an integrated global energy system, utilizing solar energy, hydrogen energy, and other safe and sustainable sources of energy;

17.3.12.4 Push forward a global program for agricultural production to achieve maximum sustained yield under conditions which are ecologically sound;

17.3.12.5 Establish conditions for free trade within the Federation of Earth;

17.3.12.6 Call for and find ways to implement a moratorium on nuclear energy projects until all problems are solved concerning safety, disposal of toxic wastes and the dangers of use or diversion of materials for the production of nuclear weapons;

17.3.12.7 Outlaw and find ways to completely terminate the production of nuclear weapons and all weapons of mass destruction;

17.3.12.8 Push forward programs to assure adequate and non-polluted water supplies and clean air supplies for everybody on Earth;

17.3.12.9 Push forward a global program to conserve and re-cycle the resources of Earth.

17.3.12.10 Develop an acceptable program to bring population growth under control, especially by raising standards of living.

Section 17.4   Second Operative Stage of World Government

17.4.1 The second operative stage of World Government shall be implemented when fifty percent or more of the nations of Earth have given either preliminary or final ratification to this World Constitution, provided that fifty percent of the total population of Earth is included either within the ratifying nations or within the ratifying nations together with additional World Electoral and Administrative Districts where people have ratified the World Constitution by direct referendum.

17.4.2 The election and appointment of Members of the World Parliament to the several Houses of the World Parliament shall proceed in the same manner as specified for the first operative stage in Section 3.2, 3.3, 3.4 and 3.5 of Article 17.

17.4.3 The terms of office of the Members of the World Parliament elected or appointed for the first operative stage of World Government, shall be extended into the second operative stage unless they have already served five year terms, in which case new elections or appointments shall be arranged. The terms of holdover Members of the World Parliament into the second operative stage shall be adjusted to run concurrently with the terms of those who are newly elected at the beginning of the second operative stage.

17.4.4 The World Presidium and the Executive Cabinet shall be reconstituted or reconfirmed, as needed, at the beginning of the second operative stage of World Government.

17.4.5 The World Parliament and the World Executive shall continue to develop the organs, departments, agencies and activities which are already underway from the first operative stage of World Government, with such amendments as deemed necessary; and shall proceed to establish and develop all other organs and major departments and agencies of the World Government to the extent deemed feasible during the second operative stage.

17.4.6 All nations joining the Federation of Earth to compose the second operative stage of World Government, shall immediately transfer all weapons of mass destruction and all other military weapons and equipment to the World Disarmament Agency, which shall immediately immobilize such weapons and equipment and shall proceed forthwith to dismantle, convert to peacetime uses, recycle the materials thereof, or otherwise destroy such weapons and equipment. During the second operative stage, all armed forces and para-military forces of the nations which have joined the Federation of Earth shall be completely disarmed and either disbanded or converted on a voluntary basis into elements of the non-military World Service Corps.

17.4.7 Concurrently with the reduction or elimination of such weapons, equipment and other military expenditures as can be accomplished during the second operative stage of World Government, the member nations of the Federation of Earth shall pay annually to the Treasury of the World Government amounts equal to one-half of the amounts saved from their national military budgets during the last year before joining the Federation and shall continue such payments until the full operative stage of World Government is reached. The World Government shall use fifty percent of the funds thus received to finance the work and projects of the World Economic Development Organization.

17.4.8 Upon formation of the Executive Cabinet for the second operative stage, the Presidium shall issue an invitation to the General Assembly of the United Nations Organization and to each of the specialized agencies of the United Nations, as well as to other useful international agencies, to transfer personnel, facilities, equipment, resources and allegiance to the Federation of Earth and to the World Government thereof. The agencies and functions of the United Nations Organization and of its specialized agencies and of other international agencies which may be thus transferred, shall be reconstituted as needed and integrated into the several organs, departments, offices and agencies of the World Government.

17.4.9 Near the beginning of the second operative stage, the Presidium in consultation with the Executive cabinet, shall formulate and put forward a proposed program for solving the most urgent world problems currently confronting the people of Earth.

17.4.10 The World Parliament shall proceed with legislation necessary for implementing a complete program for solving the current urgent world problems.

17.4.11 The World Parliament and the World Executive working together shall develop through the several organs, departments and agencies of the World Government whatever means shall seem appropriate and feasible to implement legislation for solving world problems; and in particular shall take certain decisive actions for the welfare of all people on Earth, including but not limited to the following:

17.4.11.1 Declaring all oceans, seas and canals having supra-national character (but not including inland seas traditionally belonging to particular nations) from twenty kilometers offshore, and all the seabeds thereof, to be under the ownership of the Federation of Earth as the common heritage of humanity, and subject to the control and management of the World Government.

17.4.11.2 Declare the polar caps and surrounding polar areas, including the continent of Antartica but not areas which are traditionally a part of particular nations, to be world territory owned by the Federation of Earth as the common heritage of humanity, and subject to control and management by the World Government.

17.4.11.3 Outlaw the possession, stockpiling, sale and use of all nuclear weapons, all weapons of mass destruction, and all other military weapons and equipment.

17.4.11.4 Establish an ever-normal grainery and food supply system for the people of Earth.

17.4.11.5 Develop and carry forward insofar as feasible all actions defined under Sec. 3.10, and 3.12 of the First Operative Stage.

Section 17.5   Full Operative Stage of World Government

17.5.1 The full operative stage of World Government shall be implemented when this World Constitution is given either preliminary or final ratification by meeting either condition (17.5.1.1) or (17.5.1.2):

17.5.1.1 Ratification by eighty percent or more of the nations of Earth comprising at least ninety percent of the population of Earth; or

17.5.1.2 Ratification which includes ninety percent of Earth's total population, either within ratifying nations or within ratifying nations together with additional World Electoral and Administrative Districts where ratification by direct referendum has been accomplished, as provided in Article 17, Section 1.

17.5.2 When the full operative stage of World Government is reached, the following conditions shall be implemented:

17.5.2.1 Elections for Members of the House of Peoples shall be conducted in all World Electoral and Administrative Districts where elections have not already taken place; and Members of the House of Nations shall be elected or appointed by the national legislatures or national governments in all nations where this has not already been accomplished.

17.5.2.2 The terms of office for Members of the House of Peoples and of the House of Nations serving during the second operative stage, shall be continued into the full operative stage, except for those who have already served five years, in which case elections shall be held or appointments made as required.

17.5.2.3 The terms of office for all holdover Members of the House of Peoples and of the House of Nations who have served less than five years, shall be adjusted to run concurrently with those Members of the World Parliament whose terms are beginning with the full operative stage.

17.5.2.4 The second 100 Members of the House of Counsellors shall be elected according to the procedure specified in Section 5 of Article 5. The terms of office for holdover Members of the House of Counsellors shall run five more years after the beginning of the full operative stage, while those beginning their terms with the full operative stage shall serve ten years.

17.5.2.5 The Presidium and the Executive Cabinet shall be reconstituted in accordance with the provisions of Article 6.

17.5.2.6 All organs of the World Government shall be made fully operative, and shall be fully developed for the effective administration and implementation of world legislation, world law and the provisions of this World Constitution.

17.5.2.7 All nations which have not already done so shall immediately transfer all military weapons and equipment to the World Disarmament Agency, which shall immediately immobilize all such weapons and shall proceed forthwith to dismantle, convert to peaceful usage, recycle the materials thereof, or otherwise to destroy such weapons and equipment.

17.5.2.8 All armies and military forces of every kind shall be completely disarmed, and either disbanded or converted and integrated on a voluntary basis into the nonmilitary World Service Corps.

17.5.2.9 All viable agencies of the United Nations Organization and other viable international agencies established among national governments, together with their personnel, facilities and resources, shall be transferred to the World Government and reconstituted and integrated as may be useful into the organs, departments, offices, institutes, commissions, bureaus and agencies of the World Government.

17.5.2.10 The World Parliament and the World Executive shall continue to develop the activities and projects which are already underway from the second operative stage of World Government, with such amendments as deemed necessary; and shall proceed with a complete and full scale program to solve world problems and serve the welfare of all people on Earth, in accordance with the provisions of this World Constitution.

Section 17.6 Costs of Ratification

The work and costs of private Citizens of Earth for the achievement of a ratified Constitution for the Federation of Earth, are recognized as legitimate costs for the establishment of constitutional world government by which present and future generations will benefit, and shall be repaid double the original amount by the World Financial Administration of the World Government when it becomes operational after 25 countries have ratified this Constitution for the Federation of Earth. Repayment specifically includes contributions to the World Government Funding Corporation and other costs and expenses recognized by standards and procedures to be established by the World Financial Administration.

Article 18 Amendments

Section 18.1

Following completion of the first operative stage of World Government, amendments to this World Constitution may be proposed for consideration in two ways:

18.1.1 By a simple majority vote of any House of the World Parliament.

18.1.2 By petitions signed by a total of 200,000 persons eligible to vote in world elections from a total of at least twenty World Electoral and Administrative Districts where the World Constitution has received final ratification.

Section 18.2

Passage of any amendment proposed by a House of the World Parliament shall require an absolute two-thirds majority vote of each of the three Houses of the World Parliament voting separately.

Section 18.3

An amendment proposed by popular petition shall first require a simple majority vote of the House of Peoples, which shall be obliged to take a vote upon the proposed amendment. Passage of the amendment shall then require an absolute two-thirds majority vote of each of the three Houses of the World Parliament voting separately.

Section 18.4

Periodically, but no later than ten years after first convening the World Parliament for the First Operative Stage of World Government, and every 20 years thereafter, the Members of the World Parliament shall meet in special session comprising a Constitutional Convention to conduct a review of this World Constitution to consider and propose possible amendments, which shall then require action as specified in Clause 2 of Article 18 for passage.

Section 18.5

If the First Operative Stage of World Government is not reached by the year 1995, then the Provisional World Parliament, as provided under Article 19, may convene another session of the World Constituent Assembly to review the Constitution for the Federation of Earth and consider possible amendments according to procedure established by the Provisional World Parliament.

Section 18.6

Except by following the amendment procedures specified herein, no part of this World Constitution may be set aside, suspended or subverted, neither for emergencies nor caprice nor convenience.

Article 19   Provisional World Government

Section 19.1   Actions to be Taken by the World Constituent Assembly

Upon adoption of the World Constitution by the World Constituent Assembly, the Assembly and such continuing agency or agencies as it shall designate shall do the following, without being limited thereto:

19.1.1 Issue a Call to all Nations, communities and people of Earth to ratify this World Constitution for World Government.

19.1.2 Establish the following preparatory commissions:

19.1.2.1 Ratification Commission.

19.1.2.2 World Elections Commission.

19.1.2.3 World Development Commission.

19.1.2.4 World Disarmament Commission.

19.1.2.5 World Problems Commission.

19.1.2.6 Nominating Commission.

19.1.2.7 Finance Commission.

19.1.2.8 Peace Research and Education Commission.

19.1.2.9 Special commissions on each of several of the most urgent world problems.

19.1.2.10 Such other commissions as may be deemed desirable in order to proceed with the Provisional World Government.

19.1.3 Convene Sessions of a Provisional World Parliament when feasible under the following conditions:

19.1.3.1 Seek the commitment of 500 or more delegates to attend, representing people in 20 countries from five continents, and having credentials defined by Article 19, Section 3;

19.1.3.2 The minimum funds necessary to organize the sessions of the Provisional World Parliament are either on hand or firmly pledged.

19.1.3.3 Suitable locations are confirmed at least nine months in advance, unless emergency conditions justify shorter advance notice.

Section 19.2 Work of the Preparatory Commissions

19.2.1 The Ratification Commission shall carry out a worldwide campaign for the ratification of the World Constitution, both to obtain preliminary ratification by national governments, including national legislatures, and to obtain final ratification by people, including communities. The ratification commission shall continue its work until the full operative stage of World Government is reached.

19.2.2 The World Elections Commission shall prepare a provisional global map of World Electoral and Administrative Districts and Regions which may be revised during the first or second operative stage of World Government, and shall prepare and proceed with plans to obtain the election of Members of the World Parliament to the House of Emerging World Law 225 Peoples and to the House of Counsellors. The World Elections Commission shall in due course be converted into the World Boundaries and Elections Administration.

19.2.3 After six months, in those countries where national governments have not responded favorable to the ratification call, the Ratification Commission and the World Elections Commission may proceed jointly to accomplish both the ratification of the World Constitution by direct popular referendum and concurrently the election of Members of the World Parliament.

19.2.4 The Ratification Commission may also submit the World Constitution for ratification by universities and colleges throughout the world.

19.2.5 The World Development Commission shall prepare plans for the creation of a World Economic Development Organization to serve all nations and people ratifying the World Constitution, and in particular less developed countries, to begin functioning when the Provisional World Government is established.

19.2.6 The World Disarmament Commission shall prepare plans for the organization of a World Disarmament Agency, to begin functioning when the Provisional World Government is established.

19.2.7 The World Problems Commission shall prepare an agenda of urgent world problems, with documentation, for possible action by the Provisional World Parliament and Provisional World Government.

19.2.8 The Nominating Commission shall prepare, in advance of convening the Provisional World Parliament, a list of nominees to compose the Presidium and the Executive Cabinet for the Provisional World Government.

19.2.9 The Finance Commission shall work on ways and means for financing the Provisional World Government.

19.2.10 The several commissions on particular world problems shall work on the preparation of proposed world legislation and action on each problem, to present to the Provisional World Parliament when it convenes.

Section 19.3   Composition of the Provisional World Parliament

19.3.1 The Provisional World Parliament shall be composed of the following members:

19.3.1.1 All those who were accredited as delegates to the 1977 and 1991 Sessions of the World Constituent Assembly, as well as to any previous Session of the Provisional World Parliament, and who re-confirm their support for the Constitution for the Federation of Earth, as amended.

19.3.1.2 Persons who obtain the required number of signatures on election petitions, or who are designated by Non-Governmental Organizations which adopt approved resolutions for this purpose, or who are otherwise accredited according to terms specified in Calls which may be issued to convene particular sessions of the Provisional World Parliament.

19.3.1.3 Members of the World Parliament to the House of Peoples who are elected from World Electoral and Administrative Districts up to the time of convening the Provisional World Parliament. Members of the World Parliament elected to the House of Peoples may continue to be added to the Provisional World Parliament until the first operative stage of World Government is reached.

19.3.1.4 Members of the World Parliament to the House of Nations who are elected by national legislatures or appointed by national governments up to the time of convening the Provisional World Parliament. Members of the World Parliament to the House of Nations may continue to be added to the Provisional World Parliament until the first operative stage of World Government is reached.

19.3.1.5 Those universities and colleges which have ratified the World Constitution may nominate persons to serve as Members of the World Parliament to the House of Counsellors. The House of Peoples and House of Nations together may then elect from such nominees up to fifty Members of the World Parliament to serve in the House of Counsellors of the Provisional World Government.

19.3.2 Members of the Provisional World Parliament in categories (1) and (2) as defined above, shall serve only until the first operative stage of World Government is declared, but may be duly elected to continue as Members of the World Parliament during the first operative stage.

Section 19.4   Formation of the Provisional World Executive

19.4.1 As soon as the Provisional World Parliament next convenes, it will elect a new Presidium for the Provisional World Parliament and Provisional World Government from among the nominees submitted by the Nominating Commission.

19.4.2 Members of the Provisional World Presidium shall serve terms of three years, and may be re-elected by the Provisional World Parliament, but in any case shall serve only until the Presidium is elected under the First Operative Stage of World Government.

19.4.3 The Presidium may make additional nominations for the Executive Cabinet.

19.4.4 The Provisional World Parliament shall then elect the members of the Executive Cabinet.

19.4.5 The Presidium shall then assign ministerial posts among the members of the Executive Cabinet and of the Presidium.

19.4.6 When steps (1) through (4) of section 19.4. are completed, the Provisional World Government shall be declared in operation to serve the welfare of humanity.

Section 19.5   First Actions of the Provisional World Government

19.5.1 The Presidium, in consultation with the Executive Cabinet, the commissions on particular world problems and the World Parliament, shall define a program for action on urgent world problems.

19.5.2 The Provisional World Parliament shall go to work on the agenda of world problems, and shall take any and all actions it considers appropriate and feasible, in accordance with the provisions of this World Constitution.

19.5.3 Implementation of and compliance with the legislation enacted by the Provisional World Parliament shall be sought on a voluntary basis in return for the benefits to be realized, while strength of the Provisional World Government is being increased by the progressive ratification of the World Constitution.

19.5.4 Insofar as considered appropriate and feasible, the Provisional World Parliament and Provisional World Executive may undertake some of the actions specified under Section 3.12.of Article 17 for the first operative stage of World Government.

19.5.5 The World Economic Development Organization and the World Disarmament Agency shall be established, for correlated actions.

19.5.6 The World Parliament and the Executive Cabinet of the Provisional World Government shall proceed with the organization of other organs and agencies of the World Government on a provisional basis, insofar as considered desirable and feasible, in particular those specified under Section 3.10. of Article 17.

19.5.7 The several preparatory commissions on urgent world problems may be reconstituted as Administrative Departments of the Provisional World Government.

19.5.8 In all of its work and activities, the Provisional World Government shall function in accordance with the provisions of this Constitution for the Federation of Earth.

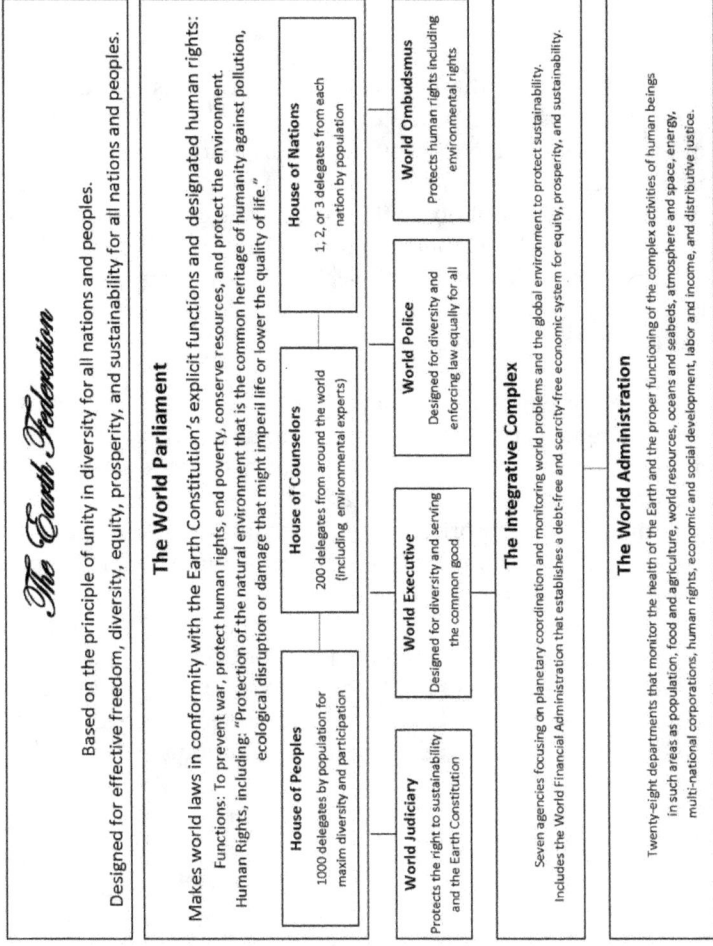

The Earth Federation

Based on the principle of unity in diversity for all nations and peoples.

Designed for effective freedom, diversity, equity, prosperity, and sustainability for all nations and peoples.

The World Parliament

Makes world laws in conformity with the Earth Constitution's explicit functions and designated human rights:

Functions: To prevent war, protect human rights, end poverty, conserve resources, and protect the environment. Human Rights, including: "Protection of the natural environment that is the common heritage of humanity against pollution, ecological disruption or damage that might imperil life or lower the quality of life."

House of Peoples
1000 delegates by population for maxim diversity and participation

House of Counselors
200 delegates from around the world (including environmental experts)

House of Nations
1, 2, or 3 delegates from each nation by population

World Judiciary
Protects the right to sustainability and the Earth Constitution

World Executive
Designed for diversity and serving the common good

World Police
Designed for diversity and enforcing law equally for all

World Ombudsmus
Protects human rights including environmental rights

The Integrative Complex

Seven agencies focusing on planetary coordination and monitoring world problems and the global environment to protect sustainability. Includes the World Financial Administration that establishes a debt-free and scarcity-free economic system for equity, prosperity, and sustainability.

The World Administration

Twenty-eight departments that monitor the health of the Earth and the proper functioning of the complex activities of human beings in such areas as population, food and agriculture, world resources, oceans and seabeds, atmosphere and space, energy, multi-national corporations, human rights, economic and social development, labor and income, and distributive justice.

W E strongly believe that the Earth Constitution is entirely adequate for the global legal renaissance of emerging world law. We citizens of the world who have encountered this marvelous document urge the rest of the citizens of our world to seriously consider the value of the Constitutions rapid adoption, so that there is the elevation of democratic self-government to the planetary level. Ratification and implementation of this Constitution will forever put an end to the war system, the poverty system, the exploitation system, and the system of environmental destruction now in place.

On the following pages are the signatures of persons from every walk of life who attended the second and fourth sessions of the World Constituent Assembly, where there were official signing ceremonies, together with a list of persons on the last page who had wanted to attend and who agreed to support the Earth Constitution. Since the time of the official ceremonies, hundreds of thousands more people have signed their support for the document.

Our immediate goal is to get both the joint ratification of a large number of national governments (about 25, perhaps) and also the measured direct ratification by the people of the world adequate to initiate the first operative stage as defined in the Earth Constitution.

Participants in the World Constituent Assembly, 16 to 29 of June, 1977, have affixed their
signatures to the draft of the CONSTITUTION FOR THE FEDERATION OF EARTH herewith:

[signatures]   India

                    MEXICO

                    EARTH, USA

Lucile W. Green   Earth, USA

                    Hon. Legal Advisor

                    Canada

T.P. Amerasinghe   Sri Lanka.

                    Kenya

Archie Casely Hayford   Ghana.

K. Koma   Botswana

Helen Tucker (Canada) Women's Universal Movement

E. L. o. Gitta   Fed. Rep. of Germany

Thane Read   U.S.A.

Spencer   India.

Rachoomuk A. Amungi   Thailand

Rose J. Chesney,   Australia.

                    Germany

                    Netherlands

Toshio Miyagake   JAPAN

ANDREA von SCHMIDT	GERMANY
Edith Kerwich	Germany
Gisela Gintzel	Germany
Klaus Frohner-Schlichtmann	Germany
Ann Mische	World, U.S.A
Gerald Mische	U.S.A.
Dr. Ludwig Chang	W. Germany
Dr. Fred Karl Scheibe	U.S.A.
Marsalger	Germany
Olga Jaeger	Germany
Beatrice Meyers	U.S.A.
Elisabeth Wassenbauer	INNSBRUCK
Theo Fenzler	Switzerland
Dr Helen K Billings	USA
Magister Kirsti Baltharsen	Finland live in Mexico
Robert Rosemond	United Peoples Federation of Earth
Valerie Hagenhuber	Austria
Herbert Gröder	"
[illegible]	"

[Page of handwritten signatures with country names]

Louis R. Gomberg	U. S. A
P. C. Malhotra	India
Hildegard Heuer	Schweiz
PURAN SINGH AZAD.	(INDIA)
Dr. Miss. Geeta Shah	INDIA.
Maria Freli	Schweiz
Karl Frenz	Türich
Bonnie Allen	U. S. A
Rustom M. Bharucha	India.
Helen Bryant	USA
Jeanne C. Burros	Unità. World
Leo J. Murray sa (Pax Christi USA)	
Simon R. Glad	Botswana
Mrs. Renée Dangoor	United Kingdom
Mr. J. Lelaka	Botswana
Peggy Lehman	Australia
Donald L. Lehman	Australia
Thomas Gibbs	AUSTRIA

Dr. Hildegard Durfee _M. D._

Hira Lynne Allen

Samar Basu India.

Robert W. Kaminski Earth USA, Allow Del

John Hoyt Holland

Yogi Shantiswaroop. India for one world

Carmel Kussman U.S.A.

Mortimer Lofaly U.S.A.

S. Hermann Weiss Austria

Kim Haroide. Canada

Ann Marin P. R

Naim Dangoor U.K.

E. Comm (Sri Lanka)

Bangalore. India

Bernadette F. Lawthers.

Craig Ann White, Ph.D. Ohio U.S.A.

Everett Rafson Wis. USA.

Mildred P. Parmelee. U.S.A.

Dr. (Mrs.) Kamoo Patel Pondichery (India)

Margaret Gadge. United Kingdom.

[handwritten signatures]   Sri Lanka

[Margaret Isely]   U.S.A.

[signature]   Austria

[signature]   PUERTO RICO

[signature]   U.S.A.

[Edward R. Reader]   Puerto Rico

[signature]   India

[Dorothy ... Barkes]   U.S.A.

[Carl F. Cattarin]   Earth!

[signature]   Denmark

[Weatherseal]   U.S.A.

[signature]   U.S.A.

[Foster, ...]   Kenya

[signature]   W. Germany

[signature]   U.S.A.

[signature]   Nigeria

[Mitsuo Miyake]   JAPAN

[signature]   Netherlands

[signature]   Botswana

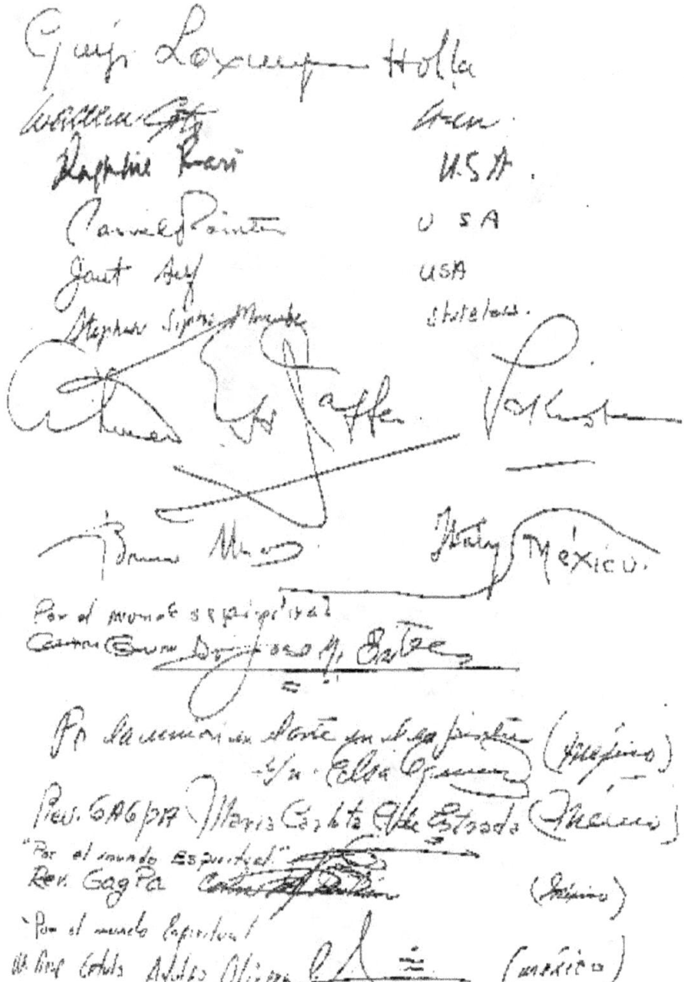

Sakoga Hage	Mexico
Jufini Holinger	Austria
[signature]	Austria
Herbert Thin	Deutschland (BRD)
Siddharth Patel	Kenya.
Khalabem Patet	India
Umesh A Patel	Great Britain.
Kamad I. Patel	Great Britain.
Ahmed Subarjo y	Indonesia
Asetyobudiart	Indonesia
Sybil Stiart	New Zealand + USA
Alice Stephens	England.
Elizabeth E. Stewart	United States
[signature]	*[signature]* Bangladesh
HARBHAJAN SINGH Khalsa	
Yogiji	USA
Sikh Dharma Western Hemp.	

Note: This list of initial signers of the CONSTITUTION FOR THE FEDERATION OF EARTH
would include several hundred more persons from fifty countries, prevented only by the cost of travel
to attend the Assembly at Innsbruck, Austria.

HELD AT TROIA, PORTUGAL, 29th APRIL, to 9th MAY, 1991

Prof. Dr. Kaiman Abraham, Hungary

Atiku Abubakar, Nigeria

Dr. Ebenezer Ade. Adenekan, Nigeria

Malcolm S. Adiseshiah, India

Abdur Rahim Ahamed, Bangladesh

Shahzada/Kabir Ahmed

Mohsin A. Alaini, Yemen

MD. Nural Alam, U.S.A.

MD. Maser Ali, Bangladesh

Dr. Terence P. Amerasinghe, Sri Lanka

Samir Amin, Senegal

Benjamin K. Amonoo, Ghana

George Anca, Romania

Mauricio Andres-Ribeiro, Brazil

Dr. Munawar A. Anees, U.S.A.

Rev. Ebenezer Annan, Ivory Coast

Jose Ayala-Lasso, Ecuador

Ir. Hasan Basri, Indonesia

Samar Basu, India

Tony Benn, United Kingdom

Prof. Mrs. Edvige Bestazzi, Italy

Petter Jakob Bjerve, Norway

Goran von Bonsdorff, Finland

Selma Brackman, U.S.A.

Jean-Marie BRETON

Jean-Marie Breton, Int. Regis. World Citizens

Tomas Bruckman, Germany (East)

Dennis Brutus, South Africa (U.S.A.)

Dr. Mihai Titus Carapancea, Romania

Prof. Henri Cartan, France

Amb. Khub Chand, India

Dr. Sripati Chandrasekhar, India

Most Rev. French Chang-Him, Seychelles

Munyaradzi Chiwashira, Zimbabwe

Dr. Pratap Chandra Chunder, India

Prof. Dr. Rodney Daniel, France

Daniel G. De Culla, Spain

Dr. Dimitrios C. Delivanis, Greece

Prof. Dr. Francis Dessart, Belgium

Raymond F. Douw, Germany

Prof. Hans-Peter Duerr, Germany

Kennedy Emekan, Nigeria

M. Necati Munir Ertekun, Cyprus

Douglas Nixon Everingham, Australia

John R. Ewbank, U.S.A.

Marjorie Ewbank, U.S.A.

Miss Lianmangi Fanal, India

Dr. Mark Farber, U.S.A.

Feng Ping-Chung, China

Prof. Dr. Mihnea Georghiu, Romania

Lucile W. Green, U.S.A.

Dr. Dauji Gupta, India

Kishaloy Gupta, India

Takeshi Haruki, Japan

Dr. Gerhard Herzberg, Canada

Jozsef Holp, Hungary

A. K. Fazlul Hogue, Bangladesh

Chowdhury Anwar Husain, Bangladesh

Margaret Isely, U.S.A. (Earth)

Philip Isely, U.S.A. (Earth)

Ram K. Jiwanmitra, Nepal

Roy E. Johnstone, Jamaica

Mohammed Kamaluddin, Bangladesh

Mohammad Rezaul Karim, Bangladesh

Rev. George Karunakaran, India

Dr. Inamullah Khan, Pakistan

Johnson S. Khan, Pakistan

Roger Kotila, Ph.D., U.S.A.

David M. Krieger, U.S.A.

Diemuth Kuebart, Germany

Jul Lag, Norway

Ben M. Leito, Netherlands Antilles

Thomas Lim, East Malaysia

Adam Lopatka, Poland

Anwarul Majid, Bangladesh

Dr. M. Sadiq Malik, Pakistan

Guy Marchand, France

Alvin M. Marks, U.S.A.

Bernardshaw Mazi, Nigeria

Dr. Zhores A. Medvedev, U. K. (USSR)

Anna Medvegey, Hungary

R. C. Mehrotra, India

Charles Mercieca, U.S.A.

Lt. Col. Pedro B. Merida, Philippines

Yerucham Meshel, Israel

Shata Mikayele, Zaire

SIGNATURES 231

Mohamed Ezzedine Mili, Switzerland

Rev. Toshio Miyake, Japan

Shettima Ali Monguno, Nigeria

Swapan Mukherjee, India

Hanna Newcombe, Canada

Brij P. Nigam, India

Josephine Okafor, Nigeria

Johnson Olatunde, Sierra Leone

Rev. Nelson Onono-Onweng, Uganda

Umit Ozturk, Turkey

Yasar Ozturk, Turkey

Linus Pauling, U.S.A.

Fernando Perez Tella, Spain

Emil Otto Peter, Austria

Dr. Alex Qualson-Sackey, Ghana

Soili Raikkonen, Finland

Sudhir Kumar Rangh, India

Thane Read, U.S.A.

Dr. Sayed Qassem Reshtia, Switzerland

Erzsebet Rethy, Hungary

Miguel B. Ricardo, Portugal

G. Rivas Mijares, Venezuela

Reinhart Ruge, Mexico

Prof. Sir A. M. Sadek, South Africa

Abdus Salam, Italy

Akber Ali Saleh, Comoros Islands

Blagovest Sendov, Bulgaria

Indira Shrestha, Nepal

Rabi Charan Shrestha, Nepal

Jon Silkin, United Kingdom

Jozef Simuth, Slovak Republic

Dr. Kewal Singh, India

Blaine Sloan, U.S.A.

Ross Smyth, Canada

Lord Donald Soper, United Kingdom

Scott Jefferson Starquester, U.S.A.

Homi J. H. Taleyarkhan, India

Rev. Yoshiaki Toeda, Japan

Dr. Duja K. Torki, Tunisia

Helen Tucker, Canada

Evelyn Utulu, Nigeria

Mrs. Justina N. Uwechue, Nigeria

Ogleva O. Uwuigbe, Nigeria

Ann Valentin, U.S.A.

T. Najat Veziroglu, U.S.A.

[signature]

Jorgen Laursen Vig, Denmark

[signature]

George Wald, U.S.A.

[signature]

Prof. D. A. Walker, United Kingdom

Kenneth R. Clark, U.S.A.

[signature]

David Daube, U.S.A.

[signature]

Njoh Ekarwakk, Cameroon

[signature]

Richard W. Wilbur, U.S.A.

[signature]

Dr. Sylwester Zawadzki, Poland

Prof. Chief J.O. Agboye, Nigeria
Dr. Francis Alexis, Grenada
Sir Abdul W.M. Ameer, Sri Lanka
Hanan Awwad, Palestine
Hon. Lukasz Balcer, Poland
Chief Dr. Kolawole Balogun, Nigeria
Dr. Sabiri O. Biobaku, Nigeria
Dr. Jur. Jan Carnogursky, Slovakia
Dr. Guvin Gerlien, Cote D'Ivoire
Amarsinh Chandhary, India
Mdm. Justice T. P. Chibesakunda, Zambia
Ashis Kumar De, India
Dr. Mostafa El Desouky, Kuwait
Dr. Rolf Edberg, Sweden
Dr. Benjamin B. Ferencz, U.S.A.
Prof. Vitalii I. Goldansky, Russia
Prof. Dr. Zbigniew Gornych, Poland
Prof. Errol E. Harris, U.S.A./U.K.
LIC Juan Horacio S., Argentina
Sir Dr. Akanu Ibiam, Nigeria
K. Jeevagathas, Sri Lanka
R. B. Junon, India
Dr. Jan Kleiner, Slovakia
Dr. Yuri A. Kosygin, Russia
Adv. Ranjan Lokhangal, India
Adv. Aqil Lodhi, Pakistan
Dr. Nikolai A. Logulciev, Russia
Mochtar Lubis, Indonesia
Perry Maison, Ghana
Kapasa Makasa, Zambia
Dr. Ignacy Malecki, Poland
Prof. Ivan Malek, Czechoslovakia
Dr. Mrs. Alla G. Massevitch, Russia
Mb1agarno S. Maanschrila, Switzerland

Dr. Mihajlo Mihajlov, Yogoslavia
Hon. Ram Niwas Mirdha, India
Dr. Robert Muller, Costa Rica
Justice M. A. Muttalib, Bangladesh
Dr. Sdeke G. Mwale, Zambia
Dr. Rashmi Mayur, India
Dr. Jayant V. Narlikar, India
Pani Nkadi, Nigeria
Osman N. Orek, Turkish Rep. N. Cyprus
Prof. Lenard Pal, Poland
Prof. Jean-Claude Pecker, France
Prof. Gamini L. Peiris, Sri Lanka
Gerard Piel, U.S.A.
Rev. Daniel O. Pepuah, Cote D'Ivoire
Prof. M. S. Rajan, India
Prof. C. N. R. Rao, India
Sri N. S. Rao, India
Michal Rusinek, Poland
Dr. Frederick Sanger, U.K.
Sir Ainsworth D. Scott, Jamaica
David Shahar, Israel
Tomu Sik, Israel
Chandan Som, India
Hon. Robert D. G. Stanbury, Canada
Dr. Bogdan Suchodolski, Poland
Abdul Lathy Sulaiman, Sri Lanka
Dr. Sol Tax, U.S.A.
Millicent Obenewaa Terry, Ghana
Dr. Walter E. Thirring, Austria
Most Rev. Desmond M. Tutu, South Africa
Kenji Urata, Japan
Dr. Pieter Van Dijk, Netherlands
Carlos Warter, M.D., U.S.A.
Rod Welford, M.L.A., Australia

A Pledge of Allegiance to the Earth Constitution

I pledge allegiance to the Constitution for the Federation of Earth, and to the Republic of free world citizens for which it stands,

One Earth Federation, protecting by law the rich diversity of the Earths citizens, One Earth Federation, protecting the precious ecology of our planet.

I pledge allegiance to the World Parliament representing all nations and peoples, and to the democratic processes by which it proceeds,

One law for the Earth, with freedom and equality for all, One standard of justice, with a bill of rights protecting each.

I pledge allegiance to the future generations protected by the Earth Constitution, And to the unity, integrity, and beauty of humankind, living in harmony on the Earth,

One Earth Federation, conceived in love, truth, and hope, with peace and prosperity for all.

Bibliography

Allott, Philip (1990). *Eunomia: New Order for a New World*. Oxford: Oxford University Press.

Armstrong, Karen (2007). *The Great Transformation: The Beginning of Our Religious Traditions*. New York: Random House.

Barber, Benjamin (1984). *Strong Democracy: Participatory Politics for a New Age*. Berkeley: U. of California Press.

Barker, Ernest (1967). *Reflections on Government*. London: Oxford University Press.

Bateson, Gregory (1972). *Steps to an Ecology of Mind*. New York: Ballantine Books.

Benzoni, Francisco J. (2007). *Ecological Ethics and the Human Soul*. Notre Dame: U. of Notre Dame Press.

Berkeley, George (1957). *A Treatise Concerning the Principles of Human Knowledge*. New York: Bobbs-Merrill Co.

Bidmead, Harold S. (1992). *A Parliament of Man: The Federation of the World*. Swimbridge, England: Patton Publications.

Biersteker, Thomas J. and Weber, Cynthia, eds. (1996). *State Sovereignty as Social Construct*. Cambridge, UK: Cambridge University Press.

Birch, Chalres & Cobb, Jr., John B. (1990). *The Liberation of Life*. Denton, TX: Environmental Ethics Books.

Blain, Bob (2004). *Weaving Golden Threads: Integrating Social Theory*. Appomattox, VA: Institute for Economic Democracy Press.

Boswell, Terry and Chase-Dunn, Christopher (2000). *The Spiral of Capitalism and Socialism: Toward Global Democracy*. Boulder, CO: Lynne Rienner Publishers.

Boucher, Douglas H. (1999). *The Paradox of Plenty: Hunger in a Bountiful World*. Oakland, CA: Food First Books.

Brecher, Jeremy and Costello, Tim (1994). *Global Village or Global Pillage: Economic Reconstruction from the Bottom Up*. Boston: South End Press.

Brinton, Crane (1948). *From Many One. The Process of Political Integration. The Problem of World Government*. Westport, CT: Greenwood Press.

Brown, Donald E. (1991). *Human Universals*. New York: McGraw-Hill.

Brown, Ellen Hodgson (2007). *Web of Debt: The Shocking Truth about Our Money System*. Baton Rouge, Louisiana: Third Millennium Press.

Brown, Lester R. (2001). *Eco-Economy: Building an Economy for the Earth.* New York: W. W. Norton & Co.

Brown, Lester R., Renner, Michael and Halweil, Brian (1999). *Vital Signs, 1999: The Environmental Trends That Are Shaping Our Future.* New York: W. W. Norton & Co.

Cairns, John Jr., (2013). http://www.johncairns.net/.

Caldicott, Helen (1994). *Nuclear Madness.* Revised Edition. New York: W. W. Norton & Co.

—. (1992). *If You Love This Planet.* New York: W.W. Norton & Co.

Capra, Fritjof (1975). *The Tao of Physics—An Exploration of the Parallels Between Modern Physics and Eastern Mysticism.* Berkeley: Shambhala.

Capra, Fritjof (1997). *The Web of Life: A New Scientific Understanding of Living Systems.* New York: Random House.

Carson, Rachel (1962). *Silent Spring.* New York: Fawcet Publishers.

Catton, Jr. William R. (1982). *Overshoot: The Ecological Basis of Revolutionary Change.* Chicago: University of Illinois Press.

Chase-Dunn, Christopher (1998). *Global Formation: Structures of World Economy.* Updated Edition. New York: Rowman & Littlefield.

Chomsky, Noam (2003). *Hegemony or Survival: Americas Quest for Global Dominance.* New York: Henry Holt & Company.

—. (1996a). *What Uncle Sam Really Wants.* Berkeley: Odonian Press.

—. (1996b). *Powers & Prospects. Reflections on Human Nature and the Social Order.* Boston: South End Press.

Chossudovsky, Michel (1999). *The Globalization of Poverty: Impacts of IMF and World Bank Reforms.* London: Zed Books LTD.

Cobb, Jr., John B. (1992). *Sustainability: Economics, Ecology & Justice.* Maryknoll, NY: Orbis Books.

Cohen, Joel E. (1996). *How Many People Can the Earth Support?* New York: W. W. Norton & Co.

Commission on Global Governance (1995). *Report of the Commission on Global Governance.* Oxford: Oxford University Press.

Constitution for the Federation of Earth (1991). Written by world citizens in four Constituent Assemblies. Lakewood, CO: World Constitution and Parliament Association. On the web at: http://www.worldproblems.net

Cook, Richard C. (2008). "Petition for a Monetary System that puts People First: Open Letter to the G-20." www.richardcook.com/articles/.

Corson, Walter H. (1990). *The Global Ecology Handbook: A Guide to Sustaining the Earths Future with the Latest Information on Air, Water, Climate Change, Energy, Toxic Waste, Tropical Forests, Population and Much More.* Boston: Beacon Press.

Daly, Herman E. (1996). *Beyond Growth: The Economics of Sustainable Development.* Boston: Beacon Press.

Daly, Herman E. and Cobb, John B. (1994). *For the Common Good: Redirecting the economy toward community, the environment, and a sustainable future.* Boston: Beacon Press.

Daly, Herman E. and Townsend, Kenneth N. (1993). *Valuing the Earth: Economics, Ecology, Ethics.* Cambridge: The MIT Press.

Descartes, Ren (1975). *The Philosophical Works of Descartes.* Vol. I. Elizabeth S. Haldane and G. R. T. Ross, trans. Cambridge: Cambridge University Press.

Dewey, John (1993). *The Political Writings.* Debra Morris and Ian Shapiro, eds. Indianapolis: Hackett Publishing Co.

Dewey, John and Tufts, James H. (1963). *Ethics—Revised Edition.* In Somerville, John and Santoni, Ronald E., *Social and Political Philosophy.* Garden City, NY: Doubleday & Company.

Diamond, Jared (2005). *Collapse: How Societies Choose to Fail or Succeed.* New York: Viking Books.

Dickens, Peter (1992). *Society and Nature Towards a Green Social Theory.* Philadelphia: Temple University Press.

Dussel, Enrique (1993). *Ethics and Community.* Robert R. Barr, trans. Maryknoll, NY: Orbis Books.

Ecimovic, Timi (2006). *The Information Theory of Nature.* Medosi, Korte, Slovenia: SEM Institute for Climate Change.

—. et al. (2007). *The Sustainable (Development) Future of Mankind.* Medosi, Korte, Slovenia: SEM Institute for Climate Change.

Edwards, Andres R. *The Sustainability Revolution: Portrait of a Paradigm Shift.* Gabriola Island, BC, Canada: New Society Publishers.

Edwards, David (1996). *Burning All Illusions. A Guide to Personal and Political Freedom.* Boston: South End Press.

Eisley, Loren (1959). *The Immense Journey: An Imaginative Naturalist Explores the Mysteries of Man and Nature.* New York: Vintage Books.

Ehrlich, Paul R. and Ehrlich, Anne H. (1991). *The Population Explosion.* New York: Simon & Schuster.

Engdahl, F. William (2009). *Full Spectrum Dominance: Totalitarian Democracy in the New World Order.* Weisbaden: edition.engdhal.

Escobar, Pepe (2006). *Globalistan: How the Globalized World Is Dissolving into Liquid War.* Ann Arbor, MI: Nimble Books.

Falk, Richard (1992). *Explorations at the Edge of Time. Prospects for World Order.* Philadelphia: Temple University Press.

—. (1993). "The Making of Global Citizenship" in *Global Visions: Beyond the New World Order.* Jeremy Brecher, John Brown Childs, and Jill Cutler, eds.Boston: South End Press, pp. 39-50.

Falk, Richard A., Johansen, Robert C., and Kim, Samuel S., eds. (1993). *The Constitutional Foundations of World Peace.* Albany: State University of New York Press.

Fromm, Erich (1947). *Man for Himself—An Inquiry into the Psychology of Ethics.* New York: Holt, Rhinehart, and Winston.

—.(1962). *Beyond the Chains of Illusion. My Encounter with Marx and Freud.* New York: Simon & Schuster.

Fromm, Erich, Suzuki, D.T., and de Martino, Richard (1960). *Psychoanalysis and Zen Buddhism.* New York: Harper & Row.

Galtung, Johan (2000). "A Structural Theory of Imperialism" in David P. Barish, *Approaches to Peace: A Reader in Peace Studies.* New York: Oxford University Press.

Gewirth, Alan (1996). *The Community of Rights.* Chicago: University of Chicago Press, 1996.

Glover, Jonathan (1999). *Humanity: A Moral History of the Twentieth Century.* New Haven: Yale University Press.

Goerner, Dyck, and Lagerroos (2008). *The New Science of Sustainability.* Chapel Hill: Triangle Center for Complex Systems.

Goldenberg, Suzanne (2013). "New Report Outlines Our Future: Climate Change Set to Make America Hotter, Drier, and More Disaster-prone." The Guardian, UK, 11 January 2013. http://www.guardian.co.uk /environment/2013/jan/11/ climate-change-america-hotter-drier-disaster?amp;buffer_share=98d2f

Gore, Al (1993). *Earth in the Balance: Ecology and the Human Spirit.* New York: Penguin Books.

Habermas, Jrgen (1998). *On the Pragmatics of Communication.* Edited by Maeve Cooke. Cambridge, MA: MIT Press.

—. (2001). *The Postnational Constellation: Political Essays.* Cambridge, MA: MIT Press.

Habicht, Max (1987). *The Abolition of War: Autobiographical Notes of a World Federalist and Collected Papers on Peace and World Federalism.* Paris: Club Humaniste.

Hardt, Michael and Negri, Antonio (2000). *Empire.* Cambridge: Harvard University Press.

Harris, Errol E. (1993). *One World or None: Prescription for Survival.* Atlantic Highlands, NJ.

Harris, Errol E. (1987). *Formal, Transcendental and Dialectical Thinking: Logic & Reality.* Albany, NY: SUNY Press.

—. (2000a). *Apocalypse and Paradigm: Science and Everyday Thinking.* Westport, CT: Praeger.

—. (2000b). *Restitution of Metaphysics.* Buffalo, NY: Prometheus Books.

—. (2005). *Earth Federation Now! Tomorrow is Too Late.* Appomattox, VA: Institute for Economic Democracy Press.

—. (2008). *Twenty-first Century Democratic Renaissance: From Plato to Neoliberalism to Planetary Democracy.* Appomattox, VA: Institute for Economic Democracy Press.

Harris, Errol E. and Yunker, James A., eds. (1999). *Toward Genuine Global Governance: Critical Reactions to "Our Global Neighborhood."* Westport, CT: Praeger.

Harris, Jonathan M., ed. (2000). *Rethinking Sustainability: Power, Knowledge, and Institutions.* Ann Arbor: University of Michigan Press.

Hegel, G.W.F. (1991). *Elements of the Philosophy of Right.* Alan Wood, ed. Cambridge: Cambridge University Press.

Heinberg, Richard (2011). *The End of Growth: Adapting to Our New Economic Reality.* Gabriola Island, BC: New Society Publishers.

Henderson, Hazel (1999). *Beyond Globalization: Shaping a Sustainable Global Economy.* West Hartford, CT: Kumarian Press.

—. (1988). *The Politics of the Solar Age: Alternatives to Economics.* Indianapolis, IN: Knowledge Systems, Inc.

Henderson and Ikeda (2002). *Planetary Citizenship: Your Values, Beliefs and Actions Can Shape a Sustainable World.* Santa Monica, CA: Middleway Press.

Hick, John (2004). *An Interpretation of Religion: Human Responses to the Transcendent.* Second Edition. New Haven: Yale University Press.

Hobbes, Thomas (1963). *Leviathan.* John Plamenatz, ed. New York: Merridian Books.

Homer-Dixon, Thomas F. (1999). *Environment, Scarcity, and Violence.* Princeton: Princeton University Press.

Hudson, Michael (2003). *Super Imperialism: The Origins and Fundamentals of U.S. World Dominance.* London: Pluto Press.

—, ed. (1996). *Merchants of Misery: How Corporate America Profits from Poverty.* Monroe, Maine: Common Courage Press.

Hume, David (1962). *On Human Nature and the Understanding.* New York: Collier Books.

Jaspers, Karl (1953). *The Origin and Goal of History.* New Haven: Yale University Press.

Jonas, Hans (1984). *The Imperative of Responsibility: In Search of an Ethics for the Technological Age.* Chicago: U. of Chicago Press.

Kant, Immanuel (1965). *Critique of Pure Reason.* Norman Kemp Smith, trans. New York: St. Martins Press.

—. (1964). *Groundwork of the Metaphysic of Morals.* H. J. Paton, trans. New York: Harper & Row.

—. (1956). *Critique of Practical Reason.* Lewis White Beck, trans. New York: Bobbs-Merrill, Inc.

—. (1957). *Perpetual Peace.* Louis White Beck, trans. New York: Macmillan.

Kafatos, Menas and Nadeau, Robert (1990). *The Conscious Universe: Part and Whole in Modern Physical Theory.* Berlin: Springer-Verlag.

Karliner, Joshua (1997). *The Corporate Planet: Ecology and Politics in the Age of Globalization.* San Francisco: Sierra Club Books.

Kitchener, Richard F. (1988). *The World View of Contemporary Physics— Does it Need a New Metaphysics?* Albany: State University of New York Press.

Klare, Michael T. (2002). *Resource Wars. The New Landscape of Global Conflict.* New York: Henry Holt & Company.

Klein, Naomi (2007). *The Shock Doctrine: The Rise of Disaster Capitalism.* New York: Henry Holt and Company.

Kohlberg, Lawrence (1984).*The Psychology of Moral Development, Volume Two: The Nature and Validity of Moral Stages.* San Francisco: Harper & Row.

Korb, Lawrence J. (2001). "10 Myths About the Defense Budget." *In These Times*, Vol. 25, No. 9, pp. 10-12.

Korten, David C. (1999). *The Post-Corporate World. Life After Capitalism.* West Hartford, CT: Kumarian Press, Inc.

—. (2001). *When Corporations Rule the World.* Second Edition. Bloomfield, CT: Kumarian Press.

Krasner, Stephen D. (1999). *Sovereignty: Organized Hypocrisy.* Princeton: Princeton University Press.

Krishnamurti, J. (1989). *Think on These Things.* New York: Harper & Row.

Kuhn, Thomas S. (1970). *The Structure of Scientific Revolutions.* Second Edition, enlarged. Chicago: University of Chicago Press.

Kulic, Slavko (2011). *Suvremenecivilizacije i culture.* Zagreb, Croatia: PROFIL.

Laszlo, Ervin (1996). *The Systems View of the World: A Holistic Vision for Our Time.* Cresskill, NJ: Hampton Press.

—. (2008). *Quantum Shift in the Global Brain: How the New Scientific Reality Can Change Us and Our World.* Rochester, VT: Inner Traditions.

Levinas, Emmanuel (1969). *Totality and Infinity. An Essay on Exteriority.* AlphonsoLingis, trans. Pittsburgh: Duquesne University Press.

—. (2006). *Humanism of the Other.* NidraPoller, trans. Chicago: U. of Illinois Press.

Lifton, Robert Jay (1993) *The Protean Self: Human Resilience in an Age of Fragmentation.* New York: Basic Books.

Lindner, Evelin (2006). *Making Enemies: Humiliation and International Conflict.* Westport, CT: Praeger Publishers.

Locke, John (1978). *An Essay Concerning Human Understanding.* A. S. Pringle-Pattison, ed. Sussex: The Harvester Press.

Lovelock, James (1991). *Healing Gaia: Practical Medicine for the Planet.* New York: Harmony Books.

Maheshvaranda, Dada (2003). *After Capitalism: Prouts Vision for a New World.* New Delhi: Proutist Universal Publications.

Marchand, Guy (1979). *One or Zero: The World Will Be Mundialist or Will Be No Longer.* Paris: Club Humaniste.

Martin, Glen T. (2008). *Ascent to Freedom: Practical and Philosophical Foundations of Democratic World Law.* Appomattox, VA: Institutre for Economic Democracy Press.

—.(2009). *Triumph of Civilization: Democracy, Nonviolence, and the Piloting of Spaceship Earth.* Appomattox, VA: Institute for Economic Democracy Press.

—.(2010). *Constitution for the Federation of Earth: With Introduction, Commentary, and Conclusion.* Appomattox, VA: Institute or Economic Democracy Press.

—.(2011). *The Earth Federation Movement: Founding a Social Contract for the People of Earth. History, Documents, Philosophical Foundations.* Appomattox, VA: Institute for Economic Democracy Press.

Mayur, Rashmi, ed. (1996). *Earth, Man, and Future: For the Renaissance Men and Women of the New Millinnium.* Mumbai, India: International Institute for Sustainable Future.

McChesney, Robert W. (1997). *Corporate Media and the Threat to Democracy.* New York: Seven Stories Press.

Morganthau, Hans (1993). *Politics Among Nations.* New York: McGraw-Hill.

Munitz, Milton K. (1986). *Cosmic Understanding: Philosophy and Science of the Universe.* Princeton: Princeton University Press.

Murphy, Gardner (1975).*Human Potentialities,* New York: The Viking Press.

Nietzsche, Friedrich (1969). *On the Genealogy of Morals and Ecce Homo,* Walter Kaufmann, trans. New York: Vintage Books.

Ornstein, Robert (1991). *The Evolution of Consciousness: Of Darwin, Freud, and Cranial Fire—The Origins of the Way We Think.* New York: Prentice Hall Press.

Ornstein, Robert and Sobel, David (1987). *The Healing Brain: Breakthough Discoveries About How the Brain Keeps Us Healthy.* New York: Simon & Schuster.

Parenti, Michael (2011). *The Face of Imperialism.* Boulder, CO: Paradigm Publishers.

Petras, James and Veltmeyer, Henry, et. al. ((2005). *Empire with Imperial-*

ism: The Globalizing Dynamics of Neo-liberal Capitalism. London: Zed Books.

Philpott, Daniel (2001). *Revolutions in Sovereignty: How Ideas Shaped Modern International Relations.* Princeton: Princeton University Press.

Pinker, Steven (1995). *The Language Instinct: How the Mind Creates Language.* New York: Harper Perennial.

Renner, Michael (1996). *Fighting for Survival: Environmental Decline, Social Conflict, and the New Age of Insecurity.* New York: W. W. Norton & Co.

Reves, Emery (1946). *The Anatomy of Peace.* New York: Harper & Brothers.

Rich, Bruce (1994). *Mortgaging the Earth. The World Bank, Environmental Impoverishment, and the Crisis of Development.* Boston: Beacon Press.

Rifkin, Jeremy (1989). *Entropy: Into the Greenhouse World.* Revised Edition. New York: Bantam Books.

Rousseau, Jean-Jacques (1947). *The Social Contract and Discourses.* G. D. H. Cole, trans. New York: E. P. Dutton & CO.

Ruge, Reinhart (2003). *Profiles of Lord Rinehart.* San Marcos, Mexico: SistemasTecnicos.

Seligson, Mitchell A. and Passe-Smith, John T. (1993). *Development and Underdevelopment: The Political Economy of Inequality.* Boulder, CO: Lynne Rienner Publishers.

Sen, Amartya (1999). *Development as Freedom.* New York: Anchor Books.

Shalom, Stephen Rosskamm (1993). *Imperial Alibis: Rationalizing U.S. Intervention After the Cold War.* Boston: South End Press.

Shannon, Thomas Richard (1989). *An Introduction to World-System Perspective.* Boulder, CO: Westview Press.

Sherover, Charles E. (1989). *Time, Freedom, and the Common Good: An Essay in Public Philosophy.* Albany: State University of New York Press.

Shiva, Vandana (2002). *Water Wars: Privatization, Pollution, and Profit.* Boston: South End Press.

—. (2001). *Protect or Plunder? Understanding Intellectual Property Rights.* London: Zed Books.

—. (2000). *Stolen Harvest: The Hijacking of the Global Food Supply.* Boston: South End Press.

Smith, B. Sidney (2012). *The Good American: A Situation Report for Citizens.* Appomattox, VA: Institute for Economic Democracy Press.

Smith, J. W. (2005). *Economic Democracy—The Political Struggle of the Twenty-first Century.* Sun City, AZ: Institute for Economic Democracy.

—.(2009). Money: *The Mirror Image of the Economy.* Sun City, AZ: Institute for Economic Democracy.

Speth, James Gustave (2005). *Red Sky at Morning: America and the Crisis of the Global Environment.* New Haven: Yale University Press.

Spinoza, Baruch (2002). Spinoza: *Theological-Political Treatise.* Samuel Shirley &Seyour Feldman, eds. New York: Hackett Publishing Co.

Spruyt, Hendrik (1994). *The Sovereign State and Its Competitors. An Analysis of Systems Change.* Princeton: Princeton University Press.

Stiglitz, Joseph E. (1994). *Whither Socialism?* Cambridge, MA: The MIT Press.

—. (2002). *Globalization and its Discontents.* New York: W. W. Norton & Co.

Swimme, Brian and Berry, Thomas (1992). *The Universe Story—From the Primordial Flaring Forth to the Ecozoic Era, A Celebration of the Unfolding of the Cosmos.* San Francisco: Harper San Francisco.

Tetalman, Jerry and Belitsos, Byron (2005). *One World Democracy: A Progressive Vision for Enforceable Global Law.* San Rafael, California: Origin Press.

Tucker, Mary Evelyn & Grim, John A. (1994). *Worldviews and Ecology: Religion, Philosophy, and the Environment.* Maryknoll, NY: Orbis Books.

Tucker, Robert C. ed. (1972). *The Marx-Engels Reader.* Second Edition. New York: W.W. Norton Co.

Turner, R. Kerry, Pearce, David & Bateman, Ian (1993). *Environmental Economics: An Elementary Introduction.* Baltimore: The Johns Hopkins University Press.

Wallerstein, Immanuel (1983). *Historical Capitalism.* London: Verso.

Ward, Barbara and Dubos, Ren (1972). *Only One Earth—The Care and Maintenance of a Small Planet.* New York: W.W. Norton.

Wilber, Ken (1998). *The Marriage of Sense and Soul: Integrating Science and Religion.* New York: Broadway Books.

Williams, Chris (2010). *Ecology and Socialism: Solutions to Capitalist Eco-*

logical Crisis. Chicago: Haymarket Books.

Young, Arthur M. (1976). *The Reflexive Universe: Evolution of Consciousness.* San Francisco: Delacorte Press.

Zarlenga, Stephen (2002). *The Lost Science of Money: The Mythology of Money* (The Story of Power). Valatie, N.Y.: American Monetary Institute.

Index

www.ingramcontent.com/pod-product-compliance
Lightning Source LLC
Chambersburg PA
CBHW062206270326
41930CB00009B/1666